AF560643

# ENZYME IN MOLECULAR BIOLOGY

ENCYCLOPAEDIA OF ENZYMOLOGY - V

# ENZYME IN MOLECULAR BIOLOGY

*By*

**Dr. Arvind N. Shukla**

*School of Studies of Zoology & Biotechnology*

*Vikram University*

*Ujjain (India)*

**DISCOVERY PUBLISHING HOUSE PVT. LTD.**

**NEW DELHI-110 002**

*Published by:*
**Namit Wasan**
**DISCOVERY PUBLISHING HOUSE PVT. LTD.**
4383/4B, Ansari Road, Darya Ganj
New Delhi-110 002 (India)
*Phone* : +91-11-23279245; 23253475; 43596065
*E-mail* : discoverybooksindia@gmail.com
discoverypublishinghouse@gmail.com
namitwasan9@gmail.com
*web* : www.discoverypublishinggroup.com

***First Published:* 2009**
***Reprinted:* 2021**

**ISBN: 978-81-8356-357-4 (Set)**

**ISBN: 978-81-8356-419-9**

**Enzyme in Molecular Biology**



*Printed at:*
Infinity Imaging Systems
Delhi

# Preface

This classic *Encyclopaedia of Enzymology* has been compiled to provide an understanding of enzymology for university degree study and teaching. It comprehensively covers theory and applications of these biological life supporting catalysts in a readable book that will serve as a bridge to more advanced and specialized areas. It not only introduces the fundamentals and biochemistry of enzymes but also provides the molecular and experimental background. This unique piece of work compiled five volumes containing wealth of information for researchers, microbiologists, and biotechnologists. This excellent title is intended mainly for students taking degree course which have a substantial biochemistry component. The chief aim of this particular book is to help student understand the concepts involved in enzymology. An attempt has been made to give a perspective of each topic, and examples are quoted where appropriate.

There can be no claim to originality except in the manner of treatment and much of the information has been obtained from the books and scientific journals available in the different libraries.

The author express his thanks to his friends and colleagues whose continue inspirations have initiated him to bring out this book.

The author is painfully aware of the shortcomings, errors and misprints that have crept in, and shall be grateful to receive suggestion for improvement of the next edition from all the readers.

The author express his gratitude to Mr. Wasan and staff of M/s Discovery Publishing House Pvt. Ltd. for their whole hearted co-operation in the publication of this book.

**Author**

# Contents

# 1

# INTRODUCTION

Mass spectrometry (MS) has become a vital enabling technology in the life sciences. This chapter summarises the fundamental aspects of MS, with reference to topics such as isotopic abundance and accurate mass and resolution. A broad and comprehensive overview of the *instrumentation, techniques*, and methods required for the analysis of biomolecules is presented. Emphasis is placed on describing the soft ionisation methods and separation techniques employed in current state of the art mass spectrometers.

As defined in a publication from the International Union of Pure and Applied Chemistry (IUPAC), MS (or *mass spectroscopy*) is "the study of systems by the formation of gaseous ions, with or without fragmentation, which are then characterised by their mass to charge ratios and relative abundances". Since the publication of the last volume in Methods in *Enzymology* reviewing MS of biomolecules, there has been a revolution in the field. Two promising novel soft ionisation methods emerging at that time were not generally available, partly because both were largely *incompatible* with the typical commercial sector mass spectrometers that were in widespread use. Although the particle *bombardment/desorption* techniques of *plasma* desorption MS (PDMS), fast atom *bombardment* (FAB), and liquid secondary ion MS (LSIMS), invented a decade earlier, had been making valuable contributions to the analysis of peptides, oligosaccharides, and other *polar* and *involatile* compounds, they were largely limited to the picomole range and thus lacked the sensitivity needed to tackle the most challenging problems. During that period when analysis of intact biological molecules such as small proteins first became possible, much research was focused on attempts to ionise ever larger molecules, many of which were standards purchased from commercial suppliers. With hindsight, simply measuring the molecular weight of a large molecule is often of limited utility, whereas digesting it chemically or enzymatically to smaller moieties and measuring the masses of even a subset of these can be very informative. Today, thanks to the maturation of soft ionisation methods and new developments in mass analysers optimised for these new *ionisation* methods, MS is established as a fundamental technology in the *biological* sciences in routine use in *numerous* laboratories worldwide. There is no doubt that it is contributing to the solution of very many *fundamental* problems in biology and *medicine*.

The selection of an optimal mass spectrometric method to tackle a particular task is rarely a *straightforward* consideration. As an example, MS is the enabling technology in the field of proteomics, which involves *protein* identification in very complex mixtures such as *cell lysates*, *tissues*, or other biological samples, as well as the identification of interacting partners, characterisation of modifications, quantitation of expression levels, and studies of non-covalent protein complexes. In most cases, this involves an initial separation step, usually by one-dimensional (1D) or two-dimensional (2D) gel *electrophoresis*, perhaps labeling with an affinity-tagged ligand, followed by proteolysis with a site-specific protease such as trypsin to generate peptides, possibly multiple further *separation/enrichment* steps, then the mass *spectrometric* analysis of the peptide mixtures. Some of the questions that arise in choosing a mass spectrometer to carry out various aspects of these tasks are as follows:

- What level of sensitivity is required?
- How accurate must molecular weight measurements be, and is there a limit to the accuracy that is required or even useful?
- Will the determination of peptide molecular weight values be sufficient or will sequence data be necessary, and if so, what are the optimal techniques?
- Is it better to separate the peptide mixtures before MS analysis, for example, by high-performance liquid chromatography (HPLC)-MS, or can sufficient information be obtained on unseparated mixtures?
- If HPLC is not used, how will the selected method be affected by impurities that may be difficult to remove?
- What is the sample throughput, and is the technique amenable to automation?

- What will it cost?

Although such questions help to narrow the choices, ultimately there will be a number of alternative solutions, each of which has individual strengths and weaknesses.

## SOME DEFINITIONS AND PRINCIPLES

In addition to the 1991 IUPAC recommendations on nomenclature and definitions, in the same year the Committee on Measurements and Standards of the American Society for Mass Spectrometry also made recommendations on standard definitions of terms in use in MS. The term MS encompasses a wide range of very different methods of analysis, each with its own unique characteristics. However, understanding some basic principles applicable to all these methods is essential and will allow the strengths and weaknesses of each to be evaluated and compared. All mass spectrometers have certain features in common: a sample introduction system, a method of creating ions from neutral sample *molecules*, a means of separating them and determining their mass-to-charge ratios, and an ion detector that may also provide a quantitative measure of the number of ions being detected within a given mass window or time period. Because the functions are frequently interrelated, these are represented in Figure elsewhere in this chapter as a series of *interconnected* boxes rather than as separate components. Thus, the method of sample introduction is intimately linked to the *ionisation* step, and the choice of mass *analysers* may also be predicated on the choice of ionisation technique. Therefore, it is difficult to discuss any of these components in *isolation*.

The primary function of a mass spectrometer is to separate charge-bearing molecules according to molecular mass and to measure their mass numbers. The species studied must be ions rather than molecules because mass separation relies on the properties of charged particles moving under the influence of electric and magnetic fields, usually in high *vacuum* where the mean free path is sufficient to ensure that they mostly travel without collisions. Ions may be positively or negatively charged, although most mass *spectrometric* experiments are performed on positive ions. Positive ions can be formed either by removal of one or more *electrons* from each molecule or by the addition of one or more cations. A charged species that is formed by the removal or attachment of an electron is referred to as a molecular ion, whereas ionisation that is achieved by addition or removal of a charged atom or molecule such as $H^{þ}$ or $NH4^{þ}$ gives a species that is termed a quasi-molecular ion. The use of an earlier expression, *pseudo-molecular* ion, is not recommended. The addition of protons is the favoured method for ion production in *biological* MS, so technically these are quasi-molecular rather than molecular ions. However, in common usage, they, too, tend to be referred to as *molecular ions*.

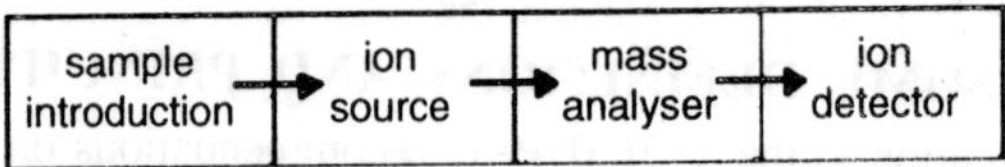


*Figure 1.1 : The basic components of a mass spectrometer.*

Ions are separated either spatially, as by a magnetic field, or temporally, as in time of flight (TOF). Even though the underlying physical characteristics of these separation methods are well understood, because of potential deviations from ideal behaviour, ion mass numbers are invariably obtained by comparison and calibration of the mass scale with known standard substances rather than by calculation from first principles. The degree of separation achieved is dependent on both the mass and the magnitude of the charge. By convention, the *abscissa* of the mass spectrum is calibrated in *m/z* units, where m is the mass and *z* is the number of electronic charges. The Thomson (Th) has been used to describe *m/z* units, although it has not achieved wide acceptance. Atomic and molecular weights are dimensionless because they are derived from a ratio of elemental mass values multiplied by 12.0000, the denominator being the mass of a $^{12}C$ atom. Nevertheless, they are frequently ascribed the symbol u, as recommended by the IUPAC, or *Daltons* (Da). An earlier *symbol* amu, an acronym for atomic mass unit, is no longer recommended. Although it is quite common to see the abscissa of a mass spectrum labeled with Da, technically this is incorrect because multiply charged ions appear not at m but at *m/z*.

**Resolving Power and Resolution**

The degree of separation achievable is determined by the resolving power (RP) of the mass *analyser*. When two peaks of mass m and m þ Lm are just resolved according a specified *criterion*, the RP is defined as m/Lm, (e.g., if ions of mass 100.0 and 100.1 can be separated, then RP 1/4 100/0.1 1/4 1000). This measurement can be made on a single peak by dividing the mass of the peak by its width without the need to observe the degree of separation. However, as shown in Figure elsewhere in this chapter, there are different criteria for ion separation: originally Lm was measured at 5% above the baseline, but more recently it has come to be measured at 50%. The first criterion is commonly called the 10% valley definition, because two equal height peaks intersecting at the 5% point overlap to give a valley equal to 10% of their height. The second, based on the full width of a peak at half its maximum height (FWHM), is now the most widely adopted definition and is used in this chapter. However, because of the addition of the ion currents where adjacent peaks overlap, the FWHM effectively defines the point at which ions of different mass can no longer be separated.

The separation of the peaks within a mass spectrum obtained by an instrument of a specific RP is expressed in terms of resolution. This is usually reported as the reciprocal of RP and is expressed in parts per million (ppm). Thus, an instrument of RP of 10,000 gives a mass spectrum in which the peaks show a resolution of 1/10,000 (i.e., 100 ppm). Many mass analysers such as a magnetic sector or a TOF have an approximately constant RP over their mass range, which means that

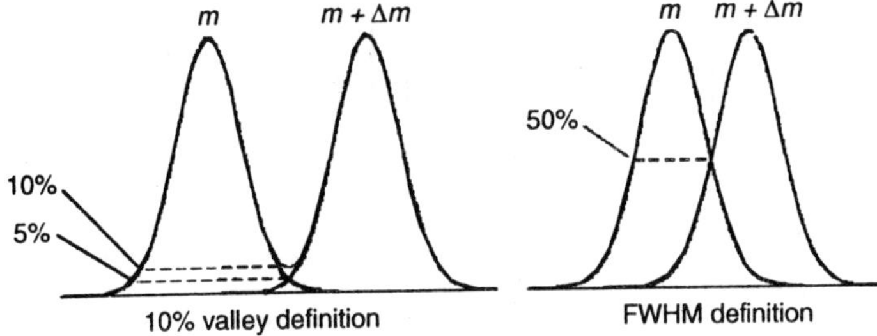


*Figure 1.2 : Criteria for alternative definitions of resolving power.*

the peak shapes become broader with increasing mass. In such an instrument, the ability to resolve ions depends only on m and is independent of m/z. Thus, if RP 1/4 1000, ions of m 1/4 1500.0 and 1501.0 will not be separated, even when they are doubly charged and appear in the mass *spectrum* at m/z 750.0 and 750.5, respectively. Other instruments such as the quadrupole analyser have an approximately constant peak width over the full mass range, in which case the definition RP 1/4 m/Lm is mass dependent, so the RP is usually expressed in m/z units instead. Thus, a quadrupole mass analyser that just separates peaks differing by 1 m/z unit is said to give a resolution of 1 mass unit. All multiply charged ions have peak spacing less than 1, so they would not be resolvable with such an analyser, independent of mass.

**Isotopes, Peak Shapes, and Mass Measurements**

The stable heavy isotopes of the elements that make up most biological molecules ($^{13}C$, $^{2}H$, 15N, $^{18}O$ , etc.) have a major impact on the appearance of ions in a mass spectrum and on the way masses are defined and measured. The molecular ions of a pure organic compound will contain a mixture of species of different *isotopic* compositions, the distribution of which is determined by the natural abundances of the various stable *isotopes* of each element. Thus, even a pure compound gives a cluster of multiple molecular peaks in the mass spectrum, each peak separated by approximately 1 Da. The first peak in the *isotopic* cluster (that of lowest m/z) is made up of the lightest isotopes of each element respectively ($^{12}C$, $^{1}H$, $^{14}N$, $^{16}O$, etc.) and is referred to as the *monoisotopic* peak. As illustrated in Figure elsewhere in this chapter for singly charged ions of the peptide hormone *bradykinin*, this gives the most intense peak for compounds of relatively low molecular

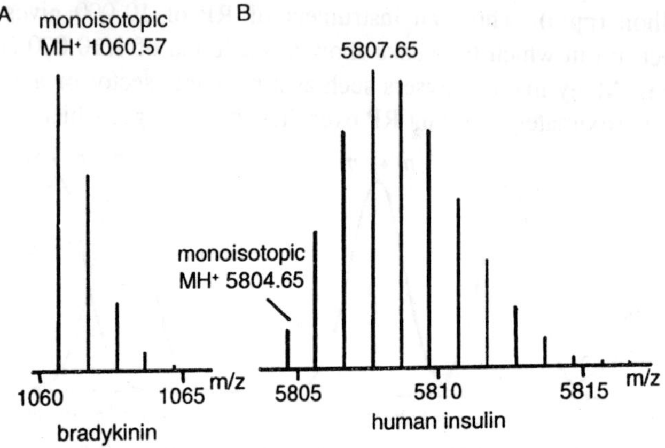


*Figure 1.3 : Calculated isotopic peak intensities for (A) bradykinin and (B) human insulin. The m/z scales use nominal mass numbers, whereas the peak positions are defined by accurate mass numbers.*

weight. However, as shown in Figure elsewhere in this chapter for the small protein insulin, as the *molecular* weight increases, the height of the second and subsequent *isotopic* peaks become greater than that of the monoisotopic peak. For bradykinin, the monoisotopic ion represents 52% of the total ensemble of molecular ions, whereas for insulin it represents only 2.6%. At the *molecular* weight of medium sized proteins (25–50 kDa), the cluster of peaks becomes broader and eventually the *monoisotopic* peak becomes vanishingly small. Many types of mass *spectrometers* can resolve the adjacent peaks in the clusters at low mass (e.g., for typical peptides of mass 500–2000 derived from a protein by tryptic digestion). However, at the molecular weight of proteins larger than *insulin*, very few instruments can resolve the isotopic clusters and they appear in the *spectrum* as a single broad peak.

Because of mass defects that arise from differences in nuclear binding energies, atomic masses do not precisely match the sum of their subatomic components, so the atomic weights of elements and their isotopes differ from integer values by small amounts. Ions of the same *nominal* mass but different *atomic composition*, termed isobars, have subtly different mass values. If these can be measured accurately, the atomic composition of the ions may be defined. A simple example is given by ions of nominal mass 16, the compositions of which could be $CH_4$ (16.0313), $NH_2$ (16.0187), or O (15.9949). The first two of these differ in mass by 1 part in 1269, so a mass spectrometer would need an RP higher than that to separate them if both ions occurred in the same spectrum. To define the composition of one of these ions would require any mass measurement error to be less half the reciprocal of the mass difference (i.e., <394 ppm). Most modern mass spectrometers would easily meet and exceed this performance. However, although such an analysis is straightforward for ions of low mass, it becomes increasingly difficult as the mass increases, partly because the absolute errors in mass measurement increase, but more so because the number of possible compositions of ions of any given mass increases nonlinearly with increasing mass.

Molecular masses may be either "*nominal*" or "*accurate*." The nominal masses are integer values based on H = 1, C = 12, N = 14, O = 16, and so on, which take no account of the fractional parts of the actual atomic masses. By contrast, the "*accurate*" masses include the fractional part, although whether experimentally measured values are truly accurate depends on several factors including the calibration of the m/z scale, which may be "*external*" or "*internal*." External calibration is carried out by comparison with the positions of peaks in a separate spectrum of a standard compound or mixture, recorded either before or after the spectrum of the sample of interest. This relies on the stability of the various components of the mass spectrometer over the time period between the recording of the calibration and the sample. This is unlikely to be as accurate as an internal *calibration* that is based on the positions of certain ions of known composition within the mass spectrum of the sample. Known trypsin autolysis fragments that commonly occur in unseparated *protein* digests are frequently used for internal calibration for the analysis of peptide mixtures. There is a connection between RP and the accuracy with which mass numbers can be measured, but they are not linked by any simple relationship. Mass measurement is based on defining the position of a peak on the m/z scale. In general, mass *spectrometers* of high RP enable this to be done more accurately; for example, with internal calibration, an instrument of RP>10,000 (resolution < 100 ppm) should be able to give mass measurements with errors <10 ppm. High RP also makes it more likely that a peak will correspond to ions of a single atomic composition, rather than multiple overlapping components of different compositions. Note that mass measurements made on *unseparated* ions of different compositions and unknown relative abundance will be uninformative or *misleading*.

The existence of stable isotopes of most commonly occurring elements gives rise to two scales for atomic masses (i.e., *monoisotopic* and *average chemical masses*). The latter is derived from weighted averages based on the masses and natural abundances of the various isotopes of each element. Thus, the *monoisotopic* atomic mass of carbon is defined as 12.0000 precisely, corresponding to the atomic mass of a $^{12}C$ atom, whereas the average mass is approximately 12.011, calculated from the natural occurrence of $^{12}C$ and $^{13}C$ in the environment, the abundance of $^{13}C$ ranging from 0.98% to 1.15%, with a typical value of approximately 1.1%. If the individual isotopic peaks are resolved for a molecular ion, *monoisotopic* atomic weights should be used to calculate the mass value of the first peak in each cluster, giving a molecular weight referred to as the monoisotopic value. When a mass spectrometer is unable to resolve the separate isotopes, average atomic masses are used to calculate *molecular* weights. For typical proteins, the difference between *monoisotopic* and average atomic weights is approximately 0.6/1000. Theoretically, the masses of noise free, well resolved *symmetrical* peaks can be defined by the positions of their peak tops. However, the typical situation is less ideal and a peak position should be defined by its *centroid*, giving an average mass for all of the ions contributing to the peak. This is particularly important for unresolved isotopic peaks, because their profiles will be asymmetrical because of uneven isotopic distributions.

Some of these points are illustrated in Figure elsewhere in this chapter, which shows simulated peak shapes for singly protonated bradykinin ions of monoisotopic mass 1060.56 at different FWHM resolutions. The simulations are based on *Gaussian* peak shapes, which are characteristic of many types of mass spectrometers. At the highest RP shown here (4000), the individual components are completely separated and the peak top coincides with the monoisotopic mass. At RP 1/4 2000, the tails of the peaks overlap and it can be seen that this is close to the 10% valley definition for the separation of peaks. However, the peaks continue to be well defined and the overlap does not affect the positions of the peak tops. At RP 1/4 1500, the peaks are extensively merged and there is a small shift of the monoisotopic peak top by 0.01 Da toward higher mass, equivalent to an error of 10 ppm. At RP 1000, the peaks are not resolved at all, although sufficient

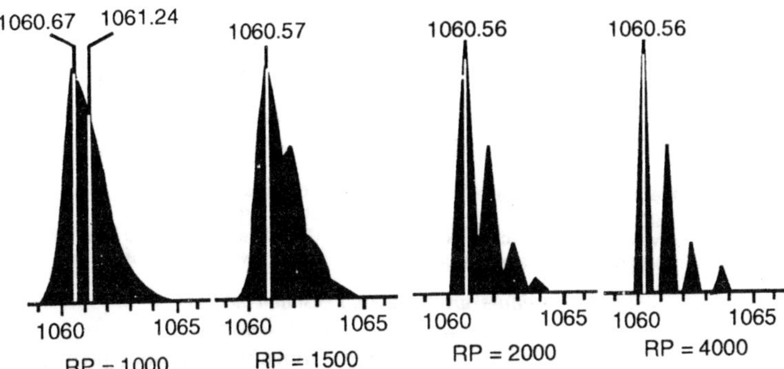


*Figure 1.4 : The effect of resolving power on simulated stable isotopic clusters for bradykinin.*

structure can be seen to indicate the underlying presence of adjacent unresolved ions. Here, the peak top is shifted to higher mass by 0.11 Da, which compared with the monoisotopic mass is an error of 100 ppm. However, comparing the peak top with the average mass of 1061.26 gives a much larger error, emphasising that it is essential to use the *centroid* rather than the peak top for average mass assignments.

**Soft Ionisation**

Until 30 years ago, mass spectrometers used in the analysis of organic compounds employed ionisation by electron impact (EI) upon isolated analyte molecules that had been evaporated into a vacuum from the liquid or solid state by heating. EI causes electronic excitation with the expulsion of an electron, giving predominantly singly charged positive ions as in Figure elsewhere in this chapter. Such ions are often described as radical cations and odd electron ions, because the removal of a single electron leaves an unpaired electron. The efficiency of ionisation is low (1 in $10^4$), but this not a problem if sufficient material is available. Heating is likely to cause decomposition of *biomolecules*, and EI transfers sufficient energy that many molecular ions undergo unimolecular dissociation to smaller fragments before analysis.

Fragment ions can be valuable because they give structural information, but compounds that contain particularly labile bonds are likely to dissociate to such an extent that no molecular ions are observed. Such an approach is not effective for the analysis of most polar, involatile, and thermally unstable *biological molecules*, although their properties may be improved by derivatisation to reduce *polarity*, increase *volatility*, and perhaps induce some characteristic fragmentation reactions.

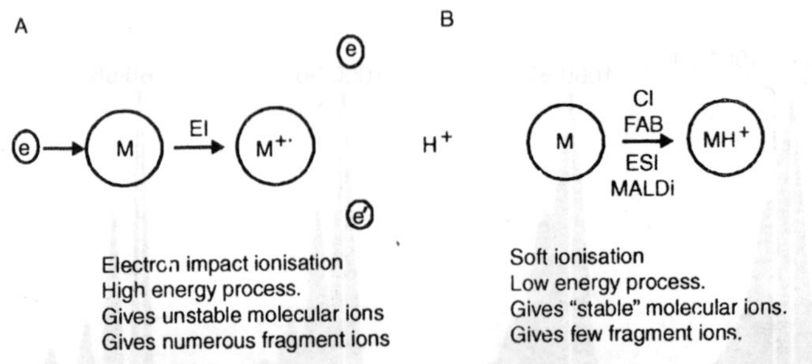


*Figure 1.5 : Mass spectrometric ionisation methods: (A) electron impact ionisation, (B) soft ionisation.*

Since about 1970, a series of alternative ionisation methods provided improvements for biological analysis, including chemical ionisation (CI) and FAB. Compared with EI, these are classified as soft ionisation methods because they transfer less vibrational energy and form ions having lower internal energies. As illustrated in Figure, they give predominantly protonated ions that have no unpaired *electrons* and are inherently more stable than the radical cations formed by loss of an electron. Two soft ionisation techniques developed in their modern form in the late 1980s proved so superior that they now account for virtually all applications of MS for the analysis of polar, involatile, and thermally labile *biomolecules*, namely *electrospray* ionisation (ESI) and matrix assisted laser desorption ionisation (MALDI). The fundamental characteristics of each meant that the choice of ionisation method largely dictated the nature of the mass *analyser*. ESI involves spraying a continuous stream of dilute solution of *analyte* from a needle held at a high potential into a chamber at *atmospheric* pressure. For positive ion analysis, a typical solvent is 50% *acetonitrile* or *methanol* acidified with 1% *acetic acid* or 0.1% *formic acid*, conditions that ensure the unfolding of proteins and extensive protonation of most basic sites. Formation of droplets and evaporation of the solvent produces a continuous stream of ions, and initially ESI was applied exclusively to mass analyders that work in a continuous mode, such as the *quadrupole*. The continuous source of ions could also be applied to magnetic sector analysers but their narrow ion beam acceptance angle and requirement for high accelerating potential made such instruments less than ideal. Although some analysis of small molecules such as the study of drug *metabolites* employs *atmospheric* pressure chemical ionisation, ESI has become the predominant ionisation method for HPLC MS in *biomolecular* analysis.

MALDI is a solid state sputtering/desorption method that produces ions by laser *bombardment* of crystals containing a small amount of analyte dispersed in a large amount of matrix. When carried out in high vacuum and at high accelerating voltage, the pulsed nature of laser radiation produces ions in pulses that are best suited to TOF analysis. Because the samples are solids, MALDI is less easily adapted to *chromatographic* interfacing, although there have been preliminary experiments with MALD! from liquid droplets. Nanoflow HPLC or capillary electrophoresis can be interfaced to MALD! using vacuum deposition of the column eluent directly onto either a moving mylar tape within the TOF mass spectrometer, or drop-by-drop deposition onto a conventional target plate for off-line processing, or even onto a compact disc for insertion into a purpose-built ion source (Krutchinsky eta!., 2001). An advantage of off-line processing of HPLC fractions by MALD! compared with on-line liquid chromatography (LC)-ES!MS is the freedom to analyse the same spot several times, obtaining further levels of information with successive *analyses*.

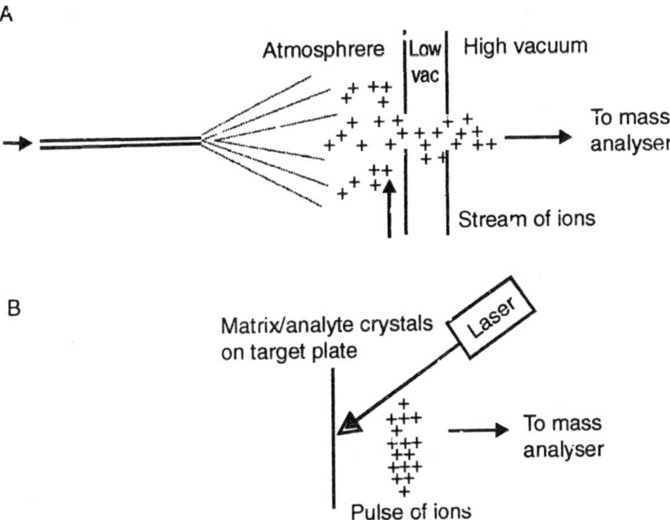


*Figure 1.6 : Alternative soft ionisation methods: (A) electrospray ionisation, (B) matrix-assisted laser desorption ionisation.*

Although both ES! and MALD! are in very widespread use, for neither technique is there full agreement on the mechanisms of ion formation and transfer to the gas phase. Overall, ES! is the simpler technique to deal with because the initial formation of ions in solution constitutes a typical example of acid-base equilibria, with pH-dependent *protonation* and dissociation reactions determining the nature of any charged species in the liquid *droplets* before evaporation. Most positive ion ESI-MS analyses are carried out in *acidic* solution, and the basic sites in an *analyte* such as a peptide effectively compete for the available protons. Peptide molecules having multiple basic sites are generally observed in the gas phase to have acquired multiple protons, suggesting that the efficiency of ion production is high. The first step in the creation of *isolated* ions is the formation of a charged cone of liquid (*Taylor cone*) from which positively charged droplets are ejected, particularly if the liquid includes an organic component that reduces its surface tension. As liquid evaporates, the droplets become smaller, and at the *Rayleigh* limit when the charge density becomes too great, they break up because of excessive repulsive forces between the positive charges. High-speed photography has established that the droplets tend to be elongated rather than spherical, and each primary droplet can emit a stream of smaller droplets that remove a disproportionate amount of charge. In the charge residue model (CRM), further evaporation causes this process to be repeated to the point at which each final droplet contains only one charged species. This is most favourable for ions of very high m/z values. By contrast, the ion evaporation model (IEM) that favours ions of lower m/z values proposes that individual ions are lost from the surface of the droplets before they reach the *Rayleigh* limit, thus reducing the net charge and preventing their *coulombic* destruction.

Despite the numerous applications of MALDI, the mechanisms of *desorption* and *ionisation* are unclear. Most commercially available MALDI instruments employ the readily available nitrogen laser that gives 3-ns length pulses at a frequency of 337.1 nm. The use of other ultraviolet (UV) lasers has been explored, but apart from differences in the absorption coefficients of *potential matrix* compounds, the spectra are largely insensitive to wavelength. UV absorption is predominantly electronic, but at the low laser *flux* typically employed in MALDI, there is no *realistic* likelihood of *multiphoton* ionisation occurring. Therefore, it must be concluded that the major function of the laser is to supply heat to the matrix. This is consistent with infrared radiation also being able to generate MALDI spectra that are similar to UV-MALDI spectra. Heating the matrix somehow effects the transfer of a portion of it into the vapor phase in the form of an expanding *plume*, presumably containing neutrals and ions and carrying the analyte. Whether the analyte was *preionised* in the solid state or becomes *ionised* in the plume, perhaps by excited state proton transfer, has yet to be established.

MALDI samples are co-crystallised with a very large molar excess of a matrix that acts as a *chromophore* for the laser radiation and protects even *thermally* sensitive analyte molecules from the direct effect of the laser irradiation. Effective matrix compounds in MALDI have been developed on a largely empirical basis, but the commonly used matrices such as 2,5-*dihydroxybenzoic* acid (DHB) and ci-cyano-4-*hydroxycinnamic* acid (ci-CHCA) popular for peptide analysis have several common features that are critical for their function. They must be *physically* compatible with the *analyte* to the degree that they

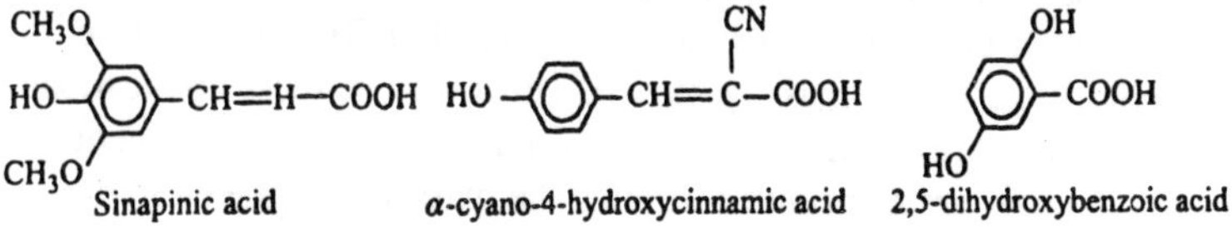

Sinapinic acid α-cyano-4-hydroxycinnamic acid 2,5-dihydroxybenzoic acid

*Figure 1.7 : Some matrices in common use for matrix-assisted laser desorption ionisation of peptides and proteins.*

can form mixed crystals, the analyte molecules being dispersed within the large excess of *matrix*. They must have an adequate coefficient for absorption of the UV radiation, which for peptides and proteins is largely achieved through the use of substituted and conjugated aromatics such as the *cinnamic* acid *derivatives*. The transfer of vibronic energy to the analyte ions is dependent on the nature of the matrix, allowing some "*fine-tuning*" of the spectra, for example, DHB imparts less internal energy than ci-CHCA and gives molecular ions that are less prone to undergo unimolecular dissociation reactions. For positive ion operation, they must act as a source of protons, so they are mostly acidic, whereas negative ion operation favours the use of basic matrices. For negative ion *oligonucleotide* analysis, basic co-matrices such as piperidine, imidazole, and triethylamine with proton affinities near or above the proton affinities of the nucleotide residues can serve as proton sinks during the *desorption/ionisation* process to enhance the ionisation (i.e., *deprotonation*) of the analyte, and they can also affect the nature and the extent of any fragmentation reactions. Matrix compounds must be relatively in-volatile and resistant to evaporation in high vacuum, a property of polar organic compounds that are extensively stabilised by hydrogen bonds. Some more or less desirable properties are relatively unpredictable; for example, the cinnamic acid derivative sinapinic acid was adopted early on as an effective matrix for *peptides* and *proteins*, but it gave rise to a *photochemical* adduct in which a sinapinic acid molecule becomes covalently attached to the analyte ion with loss of a water molecule. ci-CHCA proved to be a superior alternative because it has no hydrogen on the ci-carbon and cannot form an analogous adduct.

Compared with ESI, MALDI ionisation appears to be relatively inefficient. As stated earlier, ESI adds protons to all the basic sites in the analyte *molecules*, whereas the ions observed from MALDI are typically only singly charged. Even for a large protein containing numerous basic residues, it is rare to observe the attachment of more than two or perhaps three protons. Therefore, on statistical grounds, it is reasonable to conclude that either most a nalyte molecules remain uncharged or the ions become neutralised in MALDI and, therefore, do not contribute to the mass *spectrum.* However, in most applications, MALDI ions are formed in a high vacuum and with high kinetic en ergies, so they are transmitted through the mass spectrometer with high efficiency, whereas ESI ions are formed at atmospheric pressure and many are lost in the transfer to the vacuum system of the mass analyser. In addition, MALDI is almost always used with TOF analysis, which maximises the sensitivity of detection as it enjoys the Fellgett a dvantage, that is, as with *interferometric* detection of all wave lengths simultaneously in infrared *spectroscopy*, it allows detection of the entire ensemble of ions that are pulsed into the light tube. This contrasts with a scanni ng instrum ent that de tects only that fraction of ions passing through the mass analyser at any given time and fails to detect the much larger number that are being discriminated against. This is an important consideration when only a limited amount of sample is available. On balance, the ultimate sensitivity of the two techniques in experiments optimised for small samples proves to be surprisingly similar; for example, routine measurements on peptide mixtures are relatively straightforward at the low *femtomole* level, and specific measurements can be made at the attomole level using either method. Note that at this level of sensitivity, careful sample handling to avoid losses by ad sorption to surfaces is more likely to be the limiting factor for both *analytical* methods.

Because the technical aspects of ionisation by MALDI an d ESI are quite different, both offer advantages and disadvantages in certain situations. Unlike ESI, MALDI is relatively unaffected by modest amounts of salt and/or detergent. In fact, some detergent may be beneficial, although it may be necessary to remove excessive contaminants by cleanup of samples by solid phase extraction with "*Zip Tips*" or similar (i.e., a *pipette* tip containing a reverse phase adsorbent), allowing the peptides to be adsorbed, washed, and then eluted with a less polar solvent mixture. Also, ions formed by MALDI have higher internal energies than those from ESI, and the desorption process helps free the *analyte* ions from undesirable impurities, just as it frees them from the *matrix*. In the solution conditions of ESI, detergents may bind to and mask the sites on the analyte molecules that would normally carry a charge, and impurities are more likely to compete successfully for the available charge at the expense of the *analyte*.

It is important to note that even in the absence of impurities, analysis of peptide mixtures by either technique generally favours certain components, whereas others may be suppressed. Thus, the absence of a particular ion in the spectrum of a mixture is not proof of the absence of that peptide in the mixture. In the MALDI analysis of an unseparated tryptic digest, the peptides having less basic residues are suppressed (e.g., arginine terminated *peptides* are favoured over *lysine* terminated peptides), although chemical derivatisation as a means of reducing such discrimination has been investigated. Ion suppression in ESI MS is more dependent on polarity and surface activity. The more *hydrophobic* peptides that are less readily soluble in the liquid phase will migrate to the surface of the *droplets* and are more likely to form isolated ions in the gas phase than those that are readily soluble and remain in the bulk of the solution. This discriminatory effect in ESI can be reduced to some degree by coupling with HPLC so that at any given moment, the eluent is likely to contain no more than two or three peptides. Where possible, it is desirable to use both ionisation methods because the information obtained may be complementary. A significant advantage of using MALDI for limited quantities of sample is the ability to study the sample for as long as a sufficient portion remains on a target. As noted earlier in the context of LC MALDI, in most instances, recording a typical MALDI spectrum consumes only a small fraction of a sample, even when the total amount deposited is only a few femtomoles, allowing the target to be removed from the instrument and retained for further analysis. By contrast, ESI is a flowing technique that consumes the entire amount of a sample within the component elution time.

Note that MALDI has been used less for quantitative applications because the sample preparations are quite heterogeneous and the components of a mixture may separate during crystallisation. Thus, it is often necessary to identify a "*sweet spot*" among the *crystals* to obtain the best *spectra*. Although ESI *spectra* are generally more reproducible, ion currents are not necessarily proportional to relative concentrations in solution; therefore, quantitation requires suitable *calibration* and ideally the use of internal standards. *Employing isotopically* labeled analogs may resolve these problems for both MALDI and ESI. An example is the use of *isotopically* coded affinity tags (ICATs) developed by Aebersold. The original ICAT reagent was available in D0 and D8 forms to differentially label cysteine residues in two populations of proteins. These were then mixed and proteolysed, and the tagged peptides were extracted by avidin affinity chromatography. MS analysis of the peptide mixtures using either MALDI or ESI revealed pairs of peaks separated by 8 Da, the relative intensities of which gave a direct measure of the relative proportions of the proteins from which they were derived. Newer ICAT reagents have employed $^{13}C$ as the heavy isotope because this does not cause the isotopically differentiated peptide pairs to separate in reversed phase HPLC, unlike *deuterium*, and *chemically* cleavable links may be incorporated to facilitate the removal of the heavy biotin affinity tag. An alternative approach to chemical labeling is to carry out stable isotope labeling in cell culture (SILAC) using an *isotopically* labeled amino acid such as *arginine*, which will tag every Arg containing peptide in a tryptic digest. Although MALDI has been used to monitor *biophysical* characteristics of proteins, solution based ESI is generally favoured to observe the formation of non covalent complexes of proteins with small ligands, metal ions, DNA, and even very large multiple protein complexes.

## MALDI TOF MASS SPECTROMETERS

MALDI forms predominantly singly charged ions, so for the analysis of large molecules it requires a *mass separation* method having a high m/z range, such as a TOF analyser. Samples are dissolved in water and mixed with a matrix dissolved in water alone or water mixed with an organic solvent such as methanol or *acetonitrile*, and sometimes acidified with *formic acid* or *trifluoroacetic acid*. A small drop of 1 $\mu$l or less is deposited on a multiposition sample plate and allowed to dry. Each sample spot will contain 1–5 $\mu$g of matrix, and for a typical unseparated digest of a protein, something between a *femtomole* and a *picomole* per peptide (i.e., a molar range of peptide:matrix in the region of 1:$10^4$–$10^7$. The sample plate is loaded into the instrument through a *vacuum* lock without breaking *vacuum*. This can be viewed inside the vacuum chamber via a video camera, and its position can be manipulated as each sample is irradiated. The sample spot is normally larger than the focused laser beam and different regions of the spot can be selected manually. For *heterogeneous* sample preparations, it may be possible to select the regions of the target yielding higher ion current (i.e., the sweet spots). However, *protocols* are available that give relatively *homogeneous* sample preparations that are preferable for automated data collection, in which the laser is moved around the surface, either systematically or randomly.

Mass analysis by TOF has a long history, a commercial instrument being manufactured by the Bendix Corporation more than 40 years ago. The principle of linear TOF is straightforward, ions being pulsed into a flight tube by an accelerating potential, resulting in ion velocities that are inversely proportional to the square root of m/z. Thus, they become separated in the flight tube and their time of arrival at a detector situated at the opposite end can be converted to mass. However, the performance of the early instruments was very modest, because of variations in initial ion *energies* and ion positions before pulsing, compounded by the limitations of the *electronics* and *timing circuitry*. TOF analysis enjoyed a renaissance with the advent of MALDI, in which ions are formed in a pulse, the length of which is limited by the short pulse of the ionising laser. Furthermore, by holding the MALDI target plate at an elevated potential, the ions are instantaneously accelerated toward a grounded aperture, so all ions are pulsed into the flight tube in a narrow burst. This development occurring in the late 1980s coincided with the availability of much faster and more reliable digital electronics for *pulsing* and timing, allowing the flight times to be measured with a precision and accuracy previously unattainable. For a nitrogen laser, the pulses have a half-width of 3 ns, so all ions are formed within this short period. Typical flight times are tens of *microseconds*, so the length of the initial pulse is relatively insignificant.

MALDI made possible the ionisation of large *biopolymers* such as proteins and DNA or RNA fragments up to 100-mers, with TOF analysis being particularly suited to heavier species because it has no *theoretical* upper mass limit. Nevertheless, the *typical* RP of a few hundred for early MALDI-TOF mass spectrometers was inadequate and prevented accurate measurement of ion mass values. Two developments led to enhanced performance, namely *delayed* extraction (DE), described by a number of authors in 1995 and the general adoption for MALDI-TOF instruments of the *reflectron* or *electrostatic* ion mirror that had been described many years earlier. In the initial MALDI-TOF ion sources without DE, ions were accelerated as soon as they were desorbed by the laser pulse, with this acceleration occurring in a relatively high-pressure plume of desorbed ions and neutrals. This resulted in scattering and dissociation of many ions during acceleration, with consequent broadening of the mass spectral peaks. DE is achieved by placing a grid in front of the MALDI target and applying a high potential to both the target and the grid so the initial desorption occurs in a field-free region. In general, the ions of interest are heavy and slow moving, whereas most of the neutrals are relatively light molecules formed by thermolysis of the matrix and can diffuse away rapidly. After a delay period of perhaps 100 ns, the grid voltage is instantaneously reduced, causing the ions to be accelerated into the flight tube of the mass spectrometer, experiencing many fewer collisions. The improvement in resolution obtained from the use of DE is illustrated in Figure elsewhere in this chapter for the molecular ion of *angiotensin*, recorded in *linear* mode.

The DE source also has a focusing effect for ions with different initial velocities, as does the *reflectron* or *ion mirror*. In a linear TOF, ions of a single mass accelerated through the same potential should all have the same flight time and should arrive at the detector simultaneously. In practice, the kinetic energy imparted by the accelerating voltage is *superimposed* upon a range of energies arising from the laser desorption process.

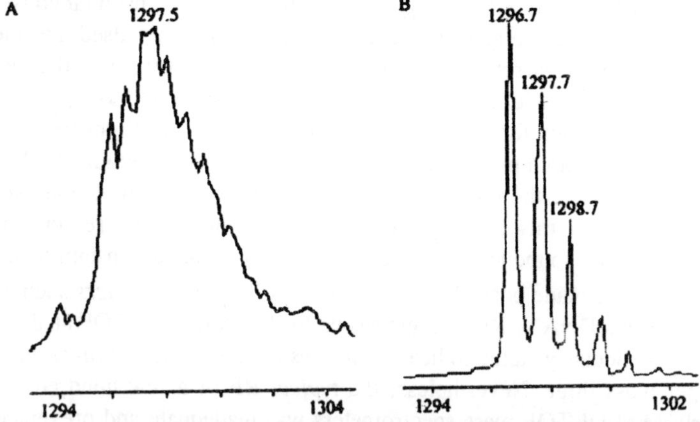


*Figure 1.8 : Matrix-assistea laser desorption ionisation-time-of-flight analysis of angiotensin ions recorded in linear mode using (A) continuous extraction and (B) delayed extraction.*

Consequently, ions of a single mass have a distribution of arrival times at the *detector*. A reflectron situated at the far end of the flight tube provides a linear potential gradient from ground to just above the accelerating voltage. This slows the ions and then reverses their direction of flight, accelerating them back toward a detector situated at a slightly offset angle. The fastest ions penetrate further into the *reflectron* and travel a longer distance before reaching the *detector* than the slowest ions. Through a suitable choice of *geometry*, it is possible to have all ions of each unique mass arriving at the detector simultaneously, even though they have slightly different velocities. With the combination of the DE source and the reflectron, modern high-performance MALDI-TOF instruments easily attain an RP of 10,000 or more, which with suitable calibration allows mass measurements to be made with an accuracy of better than 10 ppm.

The layout of a modern high-performance instrument is illustrated in Figure elsewhere in this chapter. Most such instruments have two detectors; with the reflectron voltages turned off the so-called linear detector may be used to detect heavier species such as intact protein ions. Such species may be successfully transmitted by the reflectron, but in general, they do not benefit from higher resolution measurements but may benefit from the higher sensitivity that can be achieved without the reflectron. The reflectron is predominantly used to analyse smaller species such as peptides, for which increased resolution and more

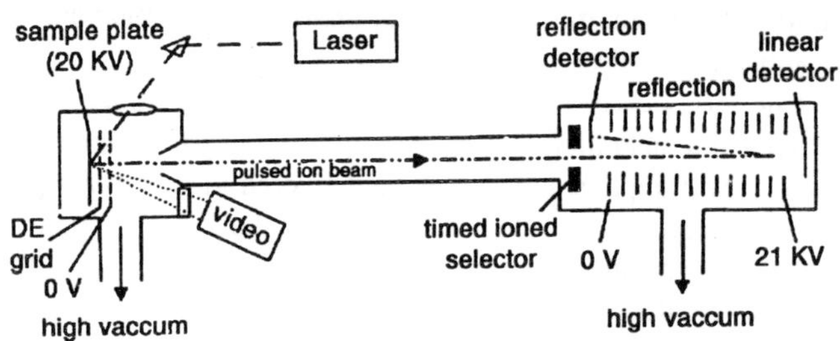


*Figure 1.9 : A high performance matrix assisted laser desorption ionisation-time of flight mass spectrometer.*

accurate mass measurements represent a substantial advantage. To optimise the efficiency of ion transmission, there may be a *guidewire* running along the center of the flight path. As ions enter the flight tube, a substantial fraction is likely to be *diverging* and *potentially* could be lost from the beam before detection. A small potential of opposite polarity to the charge on the ions applied to the guidewire makes these ions spiral around the wire as they travel down the tube, thereby maximising the fraction of ions detected. However, capturing the more divergent ions tends to *degrade* the RP, so there is a trade off between sensitivity and resolution, consequently, the *guidewire* may be replaced by focusing lenses. The low mass ions formed by the matrix (i.e., the fastest moving ions) are often abundant and they may saturate and temporarily degrade the detection system, but they provide little useful information. A timed ion selector can be used to deflect these out of the beam so that no ions are detected or recorded below a preset mass.

## MASS SPECTROMETERS OPTIMISED FOR ESI

From both a semantic and mechanistic point of view, the classification of ESI as an ionisation method for MS is questionable. More accurately, it is a means of *isolating* preexisting ions in solution and transferring them to the *vacuum* system of the mass spectrometer. Because it merely requires the evaporation of the solvent from droplets of liquid that are sprayed from a needle at *atmospheric* pressure, there is no requirement to impart significant amounts of internal energy to the ions. *Evaporative* cooling of the *shrinking droplets* is compensated for by gentle collisions with the high pressure gas, which in some source designs is heated. However, as the ions are drawn through the mass spectrometer interface into regions of lower pressure, they are accelerated by potentials that may cause more energetic collisions with the remaining air molecules, converting kinetic energy into *vibrational* energy. The *exploitation* of this feature for structural *analysis* and compound class identification is described below.

Various forms of ESI allow operation over a wide range of flow rates, from a few nanoliters per minute up to hundreds of microliters per minute. The sample solution may be delivered via a *syringe* pump connected directly to a needle through which the liquid is pumped. Alternatively, the samples may be delivered directly in the eluent from an HPLC or capillary zone *electrophoresis* (CZE). For a solution containing a fixed concentration of analyte, the observed ion currents are largely independent of flow rate. Most *biomolecular* analyses are carried out with limited amounts of sample, so lower flow rates are favoured, for example, nanobore HPLC interfaced directly with ESI requires flow rates of approximately 100–300 μl/min. A version of ESI known as *nanospray* that is optimised for very small amounts of sample uses flow rates as low as 20–50 μl/min. Here, 1–3 μl of sample solution is introduced into a *quartz* capillary, one end of which has been drawn out to a very narrow tip. The application of a high potential to the tip and a small positive pressure of gas from the other end causes the liquid to spray from the tip. Using *micromanipulators* and video cameras, the operator adjusts the position of the tip with respect to the entrance aperture of the mass spectrometer. The flow rate cannot be controlled directly, but with experience, an operator may obtain good quality spectra from 1 μl of sample for as long as 1 h.

ESI is carried out at *atmospheric* pressure, and the ions are then transferred into the *vacuum* system, an arrangement that is better suited to certain instrument types than others. When it was first introduced, ES! was most often used in *conjunction* with a quadrupole mass analyser. *Quadrupole* field mass separation technology, first developed in the 1950s by Wolfgang Paul, was well established, and single and triple quadrupole instruments were in widespread use. ESI has also become the commonly used *ionisation* method for the 3D quadrupole ion trap (QIT), although MALDI has been employed as an ionisation method for the QIT and the hybrid *quadrupole* orthogonal acceleration TOF, described later in this chapter, has employed MALDI ionisation for protein discovery, identification, and structural analysis. Compared with a sector mass spectrometer, an advantage of both the linear quadrupole and the QIT for an *atmospheric* pressure *ionisation* method is the low ion kinetic energy requirement; thus, there are no high accelerating voltages. Sector instruments use high voltages and high vacuum, both of which introduce complications in interfacing with spraying liquids at atmospheric pressure. Although *quadrupoles* were originally developed for analysis of small organic molecules and the m/z range was usually limited to about 2000, an upper limit of 3000–4000 is readily achievable, which is adequate even for ESI of proteins, as proteins sprayed from acid solution commonly give a distribution of multiply charged ions within the m/z range of about 700–2000. Despite the convenience and simplicity of linear *quadrupoles*, the resolution of such instruments is low. QITs can give superior resolution over a narrow mass range, but when these instruments are scanned over their full mass range, the resolution is also limited.

Much higher resolution can be achieved with a modified version of TOF analysis. In a conventional MALDI TOF instruments, the ions generated from a static solid sample within the vacuum system are accelerated into the flight tube within *nanoseconds* of their formation. For an alternative ionisation method such as ESI, ions are generated externally and transferred into the *vacuum* system in a procedure that takes milliseconds. If these ions were to be accelerated along the central axis of the ion source by application of a pulsed potential, the new velocity would be added to their previous component of *velocity* in the same direction. Furthermore, at the moment of acceleration, the spatial distribution of ions along this axis would further broaden the distribution of arrival times for ions of a given mass, degrading the RP and making accurate mass determination impossible. However, the ion beam from an external source such as ESI can be focused and constrained such that it has very little component of motion in any direction orthogonal to its primary axis. At a point along the ion path, a section of the beam passes between two parallel electrodes, one being solid (the *pusher*) and the other a grid (the *puller*). The application of a pulsed potential between these two electrodes accelerates the ions orthogonally into a reflectron TOF. Although the accelerating voltage is typically much lower than in a MALDI TOF (e.g., 4 kV compared with 20 kV), the resolution and mass measurement accuracy are comparable. To achieve high sensitivity in an oaTOF, it is necessary to maximise the fraction of the ions *pulsed* into the TOF. This requires that the ion velocity into the orthogonal acceleration region be matched to the duty cycle of the ion pulses, ensuring that most ions are pulsed into the TOF and minimising any fraction that passes through the accelerator without experiencing acceleration. This is difficult to achieve for the whole m/z range, because ion velocities in the beam entering the accelerator are likely to vary *inversely* with mass. The maximum pulsing rate is limited by the flight time of the slowest ions through the TOF (highest m/z), otherwise, the fastest ions from the subsequent pulse (lowest m/z) might overtake them before they reach the *detect* or. However, ion trapping before pulsing into the oaTOF can increase the sensitivity for the low mass ions without compromising the performance for the high mass ions. Typically, an oaTOF is operated with a repetition rate of about 10 kHz.

ESI of *analyte* molecules that contain multiple *basic sites* gives multiply charged ions, so molecular weights are not revealed directly. However, there is usually a distribution of charge states giving rise to a series of ions in the mass *spectrum*. Because these are spaced nonlinearly, the charge states can be determined by inspection and the molecular weight can be calculated. All manufacturers of ESI instruments provide software that does this automatically by a process known as deconvolution, which displays a profile of the molecular species present. This is shown in Figure elsewhere in this chapter for a recombinant insulin like growth factor (IGF)- binding protein (IGFBP4). In this case, the deconvoluted spectrum reveals the presence of two separate species, the lower mass component being within 0.4 Da of the calculated *molecular* weight for IGFBP4 of 25,735.3 Da. The higher mass component is 16 Da heavier, most likely attributable to partial *oxidation* of *methionine* to *methionine* sulfoxide. Note that in this example the isotopic peaks are unresolved, which is typical of most mass spectra of proteins, with Fourier transform MS (FTMS) being the only technique able to resolve such species.

Ionisation by ESI is very soft and does not usually result in fragment ions, but as mentioned earlier, suitable strategies to increase internal energy to induce fragmentation are possible. Within an intermediate pressure region of the ESI interface is a plate with an aperture commonly referred to as the *skimmer* or *nozzle*. A voltage applied to this plate can be used to increase the ion kinetic energy. Collisions with neutral gas molecules result in a fraction of the *kinetic* energy being converted to *vibronic* energy, whereby otherwise stable *molecular* ions may be energised and may undergo *unimolecular* dissociation. Thus, in an HPLC-MS experiment, it is possible to use a single *quadrupole* to observe fragment ions that are characteristic of particular compound classes. For example, by *monitoring* m/z 79 for $PO_3^-$ anions throughout the HPLC separation, *phosphopeptides* may be identified in a protein digest. Similarly, sugar *oxonium* ions such as m/z 204 for N-*acetylhexosamine* may be monitored to characterise *glycopeptides*. More *sophisticated* methods for the study of fragment ions providing substantially superior performance have been developed, whatever *ionisation* method is adopted. These are considered later in the separate section on tandem MS.

***Biophysical Studies by ESI-MS***

Because thc multiply charged ions typical of ESI can be separated by an analyser designed for ions of relatively low m/z, it is not necessary to employ a ''*high-mass*'' instrument. An exception to this general rule is the study of large protein complexes under native conditions by ESI, when relatively few sites are accessible for proton attachment; consequently, the required m/z range is significantly higher and may substantially exceed 10,000. The softness of ESI is a major advantage when studying **non-covalent** complexes, because these can be quite labile. Although the typical ESI conditions are strongly denaturing, it is possible to use *pH-neutral* solutions and to include some salts and buffers in low concentration, although these are likely to degrade the performance of the instrument and necessitate frequent cleaning of the ESI interface. It is advantageous to use volatile buffers if possible, such as *ammonium* formate or *bicarbonate*, depending on the pH level required.

Using a mass *spectrometer* to monitor complexes formed in solution requires that they can be transferred to the solvent-free vacuum without dissociation. Non-covalent interactions can be subdivided into those that may be enhanced by the removal of the *dielectric*, water, and those that will be weakened. *Hydrophobic* interactions are stabilised within an ionic *environment* (i.e., in *aqueous* salt solution) and thus are generally the most labile in the solvent-free vacuum of the mass spectrometer. In contrast, interactions between metal ions and peptides or proteins are *electrostatic* in nature and are strengthened upon removal of water. This makes the study of such interactions relatively straightforward, although the occurrence of nonspecific interactions of no biological relevance must be minimised by the use of suitable controls. It has also been demonstrated that some very large protein complexes can be stabilised in the gas phase by the presence of multiple solution-derived small *molecules* and ions.

*Hydrogen–deuterium* exchange reactions occurring in solution and monitored by ESI-MS can be the basis for other biophysical studies. Of the many *hydrogen* atoms in a protein, those that are bonded to *carbon* are generally stable, whereas OH, NH, and SH bonds are *polar* and can exchange in aqueous solution. However, the amide NH groups involved in maintaining secondary structure are more stable and less prone to exchange. For example, the *geometry* of the cit-helix requires each *backbone carbonyl* group to form a *hydrogen* bond with the spatially adjacent NH group of the amino acid on the next turn of the helix, four residues toward the C-terminus. Such effects are revealed by dissolving the protein in $D_2O$ rather than $H_2O$, the rate at which deuterium exchanges with the hydrogen atoms in the amide bonds of the protein backbone being indicative of secondary structure maintained by hydrogen bonding. Furthermore, deeply buried *hydrophobic* sections of a protein are less accessible to water, so the rate of H/D exchange in such regions will be slower than for solvent-exposed regions. Consequently, increased or decreased rates of isotopic exchange may correlate with *tertiary* structural differences.

Using nuclear *magnetic* resonance (NMR) *imaging*, *pioneering* work in this area has been carried out by Walter Englander for more than 20 years, and this method continues to enjoy widespread use. However, an alternative is to employ MS to observe changes in *molecular* mass caused by the introduction of the heavy isotope. Not only can this technique monitor changes in the molecular weight of the intact *protein*, but the protein can also be rapidly digested to smaller peptides that are then analysed by MS, usually using HPLC-MS, and possibly tandem MS. HPLC-MS is viable because H/D exchange is *catalysed* by hydrogen ions at low pH (pH $< 2$) but by *hydroxide* ions at higher pH levels ($>3$). Thus, for the pH region most likely to be of interest in protein folding, pH 3–8, the rate of H/D exchange increases by one order of magnitude per pH unit. For example, the rate of exchange in an experiment conducted at pH 7 can be slowed by a factor of 10,000 by reducing the pH to 3, which is close to *optimum* for digestion by the nonspecific protease pepsin and is the pH used for reversed phase HPLC separations. The rate may be reduced further if the temperature is lowered from ambient to close to 0°. Thus, further exchange is effectively "*frozen*" as soon as the pH and temperature are reduced. By digesting the protein into relatively small peptides, it is possible to isolate and identify the regions that show the slowest exchange and thereby determine those residues involved in maintaining the structure.

ESI MS requires only extremely small amounts of material for such experiments (e.g., a few *microliters* of *micromolar* solution), whereas NMR typically requires 0.5 mi of *millimolar* solution. Other advantages over NMR include the relatively simple assignment of peptide masses to defined regions of the sequence, whereas NMR requires a detailed assignment of each of the amide protons. Further, the mass spectrometric method allows experiments to be carried out rapidly over a wide range of pH levels. In a refinement of the mass spectrometric method, it is possible to use *collision* induced dissociation (CID) and tandem MS to define the degree of exchange at individual residues within a *peptide*.

## TANDEM MASS SPECTROMETRY

Early success with protein identification by automated database searching was based on mass mapping of peptides derived by tryptic digestion of the protein, usually separated by 2D sodium *dodecylsulfate* (SDS) *polyacrylamide* gel electrophoresis (PAGE) and analysed by MALDI TOF. In 1993, several groups independently demonstrated that a MALDI mass map of the peptides in an unseparated protein digest could correctly identify a protein that was present in a database, even when only a subset of the peptides was ionised and detected. Because every protein can potentially give a unique pattern of *peptides*, this technique is often referred to as peptide mass *fingerprinting* (PMF). However, the uniqueness of the match depends on the fraction of potential *peptides* that is actually observed and the accuracy with which their masses can be measured. Because of the high level of activity in genomic sequencing and the completion of several *genomes*, databases have grown dramatically larger, enforcing a requirement for more accurate mass measurement to increase the selectivity and specificity of mass mapping. So now PMF is rarely adequate and peptide sequence determination is virtually essential. This is certainly the case for protein identification in complex mixtures when there are too many peptide signals and PMF returns an unacceptably high number of candidates; to obtain protein sequence when no match is found in a database; or to identify and *localise* posttranslational modifications. Sequence information can be obtained readily by tandem MS, frequently known as MS/MS, which is illustrated schematically in Figure elsewhere in this chapter. Conventional MS produces ions that are separated by m/z and analysed directly. If a soft ionisation method is used (i.e., one that causes minimal fragmentation), the mass spectrum will yield the molecular weight values of compounds present in the analyte but little or no structural information. In a tandem mass spectrometer, ions of a particular mass number selected by the first mass *analyser* (MS1) can be subjected to collisions with neutral gas atoms or molecules that cause excitation of the ions, resulting in *unimolecular* decomposition. This process is known as CID. The ionic fragments are then separated in a second analyser (MS2), yielding structural information such as the amino acid sequence of a peptide. Collision conditions can be defined for two

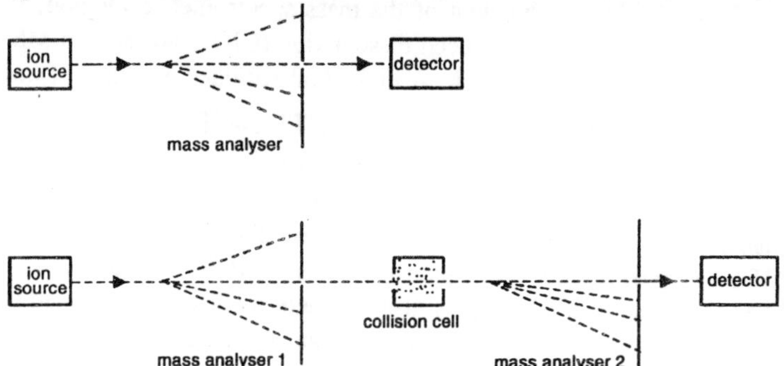


*Figure 1.10 : The principles of (A) mass spectrometry and (B) tandem mass spectrometry.*

different regimens of CID, so either the high velocity ions typical of sector instruments and MALDI TOF mostly undergo a single collision that converts sufficient kinetic energy into internal energy for reaction to occur or the slower ions typical of *quadrupoles*, QITs, and ion cyclotron resonance experience multiple collisions in a higher pressure gas cell, each of which transfers a smaller amount of energy but the aggregate effect of all collisions gives sufficient energy for reaction. It is also possible to transfer sufficient energy in the ionisation process for a substantial fraction of ions to dissociate, even using a soft ionisation technique such as MALDI, by using higher than normal laser power and using a relatively "*hot*" matrix such as ci CHCA. High energy CID can also be achieved in a MALDI-TOF instrument equipped with a collision cell.

Although MALDI and ESI are described as soft ionisation methods, each will yield ions with a distribution of internal energies. Those with the highest energies may dissociate before acceleration to appear as normal fragments in the spectrum or in the time spent between acceleration and analysis, in which case they are generally referred to as metastable ions. The initial internal energy also determines how much additional energy is needed from the collision events for the dissociation of otherwise stable ions to occur rapidly enough to yield fragment ions in the CID spectrum. It is instructive to consider the factors affecting unimolecular ion *fragmentation* in tandem MS, in which the selected parent or *precursor* ion dissociates to yield a new lighter ion with the loss of a *neutral* fragment. Such reactions are typical of all chemical processes in that they are controlled by *kinetic* and thermodynamic properties. To express this in its simplest form, the amount of internal energy of the primary ions, either with or without collisional excitation, determines whether they can undergo dissociative reactions within a certain time period, as defined by the geometry of the mass analyser. In the positive ion mode, soft ionisation methods all form ions by *cationisation*, giving inherently stable even-electron ions. This contrasts with the earlier method of ionisation by electron impact that removed an electron from each molecule to form an odd-electron radical cation. Most of the reactions of *odd-electron* ions involve *homolytic* bond breaking, in which a neutral radical is lost to form an even *electron fragment ion*. This is illustrated in reaction 1 of Figure elsewhere in this chapter, for a process that is sometimes described as ci-cleavage. (By convention, the movement of a single electron is indicated by a fishhook arrow.) For those ions possessing sufficient internal energy, these reactions that involve the dissociation of a single bond are rapid. An alternative group of reactions involves the breaking of more than one bond concurrent with the making of new bonds, thereby causing a rearrangement in which a neutral molecule is eliminated, as in reaction 2. An example of such a reaction that proceeds via a six-membered transition state is the McLafferty rearrangement, reaction 2b. The energy requirements for such reactions are often lower, but the entropic demands are higher, so the reactions are slower and may be seen to give metastable ions rather than conventional fragment ions. Because soft ionisation gives even electron ions, homolytic bond dissociation is unfavourable and most reactions of ions formed by ESI or MALDI involve *heterolytic* processes, frequently accompanied by *hydrogen* rearrangements, as shown in reaction 3. Unless the ions gain excess internal energy by a mechanism such as collisional activation, these tend to be slow processes. The rates of fast mass spectrometric reactions have been analysed using the technique of field ionisation kinetics (FIK). Field ionisation occurs in the vacuum in close proximity to a very sharp point or edge of a metal emitter, and FIK enables the study of fragmentation reactions that occur within the potential gradient derived from such an emitter.

The maximum amount of kinetic energy that can be converted to internal energy in a single collision is defined by the center of mass energy (ECM). Assuming the *velocity* of the target gas atom or molecule is negligible, the $_{ECM}$ depends on the original ion *kinetic* energy (KE) and the masses of the ion (mi) and neutral target (mn).

$$E_{CM} = KE{:}m_n = (m_i + m_n)$$

Thus, the mass of the neutral target plays a major role (e.g., for an ion of mass 1000 and KE of 1 keV): A neutral of mass 4 (*helium*) gives a theoretical maximum $E_{CM}$ of 4 eV, whereas mass 40 (*argon*) gives 40 eV. Consequently, the nature of the dissociation reactions will be affected significantly by the choice of neutral gas, and any dissociation that requires higher internal energy will benefit from a heavier target gas. In practice, the average amount of kinetic energy converted to internal energy will be less than the theoretical maximum, a phenomenon known as inelastic scattering, so a distribution of energies will be transferred by the *collisions*. Note also that conversion of kinetic energy to internal energy will result in the incident ions traveling more slowly after collisions. For ions that undergo CID, this will be seen as energy loss for the fragments, whereas fragment ions formed by metastable dissociation will not experience this energy loss. Ion fragmentation may also be accompanied by conversion of internal energy into kinetic energy, giving rise to a broadening of the kinetic energy distribution for the fragment ions, which will apply to both metastable and CID fragments. This *phenomenon* was well documented for sector instruments in experiments such as mass analysed ion kinetic energy *spectroscopy* (MIKES), in which energy loss apparently shifts the peaks to lower mass as slower ions experience greater deflection, a phenomenon that correlates with the energy requirements of the dissociation reactions. Slowing the ions down in a linear TOF would appear to make the ions heavier, but in theory a reflectron should compensate for this and the net effect is likely to be modest.

As stated earlier, collisions between ions and neutrals cause excitation of the ions with a distribution of elevated internal energies. The higher the average internal energy, the greater the number of channels that will be available for dissociation reactions, many possible reactions being accessible only to ions that are activated by high KE collisions. The differences between high- and low-energy CID of peptides are detailed elsewhere in this volume. Briefly, fragmentation at an amide (or "*peptide*") bond giving abion (*N-terminal fragment*) oray ion (*C-terminal fragment*) is a relatively low-energy process involving hydrogen rearrangements such as shown in reaction 3. Higher energy processes lead to cleavage elsewhere along the backbone, for example, between an cit-C and a carbonyl-C to give an a ion or simultaneous backbone and side-chain cleavage to gived or w ions that can distinguish *leucine* from *isoleucine*, or formation of the low mass *immonium* ions that can characterise the *amino acids* present. Thus, *metastable* ion spectra and typical low-energy CID spectra show sequence information in the form of b and y series ions, but they are *devoid* of some of the more structurally informative ions. Also, any particularly favourable amide cleavage such as at the N-terminus of a proline residue, where the secondary *amino nitrogen* can stabilise the positive charge, may dominate the *unimolecular* spectrum at the expense of other sequence ions. Consequently, if sufficient fragmentation is desired to derive the sequence from first principles, high-energy CID may be essential. On the other hand, if fragment ions are being used to search the database for a *peptide* sequence "*tag*", which is a series of fragment peaks that defines a sequence of amino acids within a peptide, low-energy CID will often suffice.

Previously, there were very few instruments other than multiple sectors capable of providing high-energy CID spectra, whereas there are numerous alternatives for low-energy CID, mostly using ES! interfaced to the triple *quadrupole*, the QIT, the *quadrupole* interfaced to an oaTOF, or the FT-ion cyclotron resonance (FT-ICR) spectrometer. QITs in particular have achieved wide acceptance for ES! with HPLC-MS and HPLC-MS/MS, taking advantage of their modest cost, fast scanning, ease of computer control, and tandem MS and $MS^n$ capabilities. However, recent developments have blurred the differentiation between instruments optimised for MALDI and ESI. MALDI ionisation at relatively high pressure (0.01–760 Torr) and low accelerating voltage (5–10 V) gives ions that become thermally equilibrated before being drawn into the higher vacuum region required for any mass analyser. Thus, the pulse nature of the ions illustrated in Figure elsewhere in this chapter becomes smeared out and more equivalent to a continuous stream of ions. The overall result is that the ESI and MALDI methods have become more interchangeable, and a number of instruments previously associated mainly with ES! are also available with MALDI as an option.

***Quadrupoles and QITs***

Both QITs and linear quadrupoles constrain ions to circular orbits under the influence of combined direct current (DC) and radiofrequency (RF) potentials. For any combination of DC and RF fields, only ions

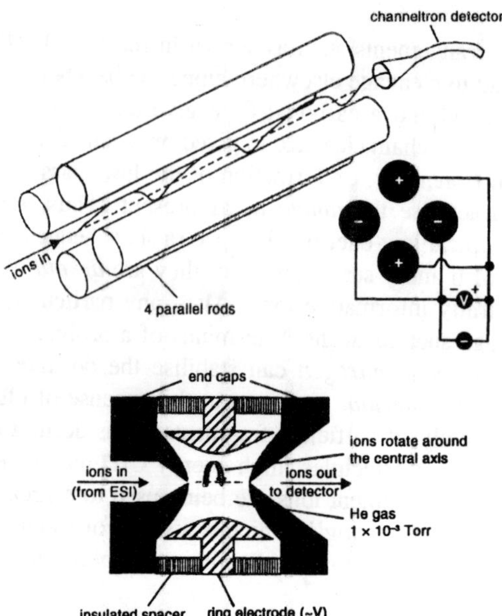


*Figure 1.11 : Two alternative designs of quadrupole mass spectrometers: (A) the linear quadrupole and (B) the quadrupole ion trap.*

of a selected m/z follow stable orbits. The fundamental difference is that ions in a 3D QIT have virtually no component of linear translation, whereas in a linear quadrupole made up of four *parallel rods*, ions travel along spiral paths, emerging to be detected or diverted into a further mass analyser. Figure elsewher in this chapter illustrates a simple linear single stage quadrupole analyser with a channeltron detector. Here, the four rods are cylindrical, although the theoretical optimum cross-section is *hyperbolic*. The detector is shown mounted on the central axis of the analyser, but improved *signal/noise* ratios can be achieved by deflecting the ions into an offset detector. There is virtually no mass selectivity in a linear quadrupole if the DC voltage is set to sero, but the RF voltage constrains ions to spiral orbits within the ion beam, so an RF-only quadrupole can make a very effective collision cell. This is used in the triple quadrupole and in a *hybrid* tandem mass spectrometer that combines a linear quadrupole for MS1, an RF-only collision cell, and an oaTOF for MS2, referred to here as a QqTOF.

Although QITs were first described about the same time as linear quadrupole analysers, for many years their commercial development lagged behind. Both linear quadrupoles and 3D QITs are relatively simple, compact, and potentially inexpensive devices that are easily controlled by computer. However, only in the last decade have QITs come to enjoy wide popularity. Now, thanks in part to a vigorous program of improvement and the realisation that the QIT could be more versatile than either the linear quadrupole or the *triple* quadrupole, they are in widespread use for a variety of analytical applications. As seen in Figure, the 3D QIT has a doughnut shaped ring *electrode* to which RF and DC voltages are applied, as well as two end caps. Ions are usually formed externally and pulsed in through an aperture in one end-cap, being constrained from further linear motion by DC potentials applied to the end-caps. They can also be pulsed out through the opposite end-cap to be detected externally. By application of appropriate voltages, ions of a single m/z can be stored within the trap, where they can be subjected to *processes* such as multiple steps of CID, often referred to as $MS^n$. Thus, despite having only a single analyser of modest proportions and cost, the QIT can be an extremely effective tandem mass *spectrometer*. The most common application of the ESI-QIT is as a mass selective detector for HPLC-MS and MS/MS, for which it was a pioneer in the development of automated CID analyses, often described as data-dependent experiments (DDEs). The simple and robust nature of the QIT makes it ideal for unattended automated operation. There are a number of variations of this, but generally, in DDE mode the instrument carries out a survey scan of the peptides eluting from the HPLC and then selects the strongest peaks for CID. After each CID experiment, the next survey scan selects new target peptides. If the same peptides as before continue to give strong signals in adjacent scans, they are excluded from repetitive analysis in favour of weaker peaks. Thus, this ensures that the maximum possible number of peptides is sampled, potentially covering a wide dynamic range. Although the QIT enables such experiments to be performed at modest cost compared with the QqTOF, its performance is inferior. Most peptides ionised by ESI give multiply charged ions, the charge states of which can be determined only if the RP of the instrument is sufficient to separate the naturally occurring *isotopes*. Unlike the QqTOF, the 3D QIT is unable to achieve this in full-scan mode. The mass range for CID is also limited, low mass ions including *immonium* ions and internal dipeptides and tripeptides not being observed. Although it is possible for proteins to be identified in a database search for sequence tags based only on ions within the mid and upper mass ranges, the lower mass ions are valuable for peptide sequencing. Note that DDE techniques are equally applicable to much higher performance tandem mass spectrometers, potentially yielding peptide *spectra* suitable for de novo sequencing.

A new addition to the QIT family is the linear 2D ion trap, which exploits elements of both the linear quadrupole and the 3D trap. A hybrid tandem version, the Q-trap, employs a linear quadrupole as MS1, an RFonly *quadrupole* as the collision cell, and the linear ion trap as MS2. Compared with the 3D QIT, this offers certain performance advantages, including less susceptibility to space charge effects and the ability to display CID fragment ions throughout the entire m/z range.

## The QqTOF

Although the oaTOF has been employed as the sole mass analyser in a number of successful home-built and commercial mass *spectrometers*, probably its true strength is as the second analyser in a hybrid tandem mass spectrometer. A sector instrument with an oaTOF designed for ESI or MALDI operation allowed high-resolution separation in both MS1 and MS2, but the reduced *sensitivity* imposed by the narrow acceptance angle of the sector analysers and technical limitations of combining ESI and MALDI with a sector instrument prevented this technique from achieving more general acceptance. However, the use of a quadrupole mass analyser as MS1 and an RF-only quadrupole collision cell in combination with an oaTOF as MS2 presents a very powerful combination in terms of sensitivity, resolution, and mass range. Illustrated in Figure elsewhere in this chapter, the QqTOF is considered by many users to be the instrument of choice for high-performance

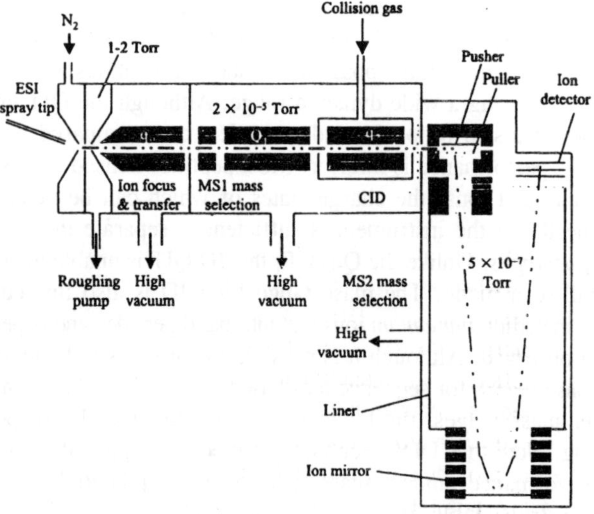


*Figure 1.12 : A Qq-time-of-flight tandem mass spectrometer.*

HPLC-MS/MS, giving resolution in the region of 10,000, mass range in MS mode of 12,000 or more (RF-only quadrupole), and mass measurement accuracy of 10 ppm or better. For MS/MS operation, the mass range is limited by the maximum m/z selectable by the quadrupole, typically 3000, but because most CID experiments for biopolymer characterisation are carried out on smaller fragments, such as peptides from a digested protein, this is generally adequate. As described earlier, to improve the duty cycle and, therefore, the sensitivity of the QqTOF for low mass ions by up to 20 times, one manufacturer has introduced a degree of mass-dependent trapping immediately before the accelerator. Ions are slowed down in a mass-selective fashion and then pulsed into the accelerator region immediately before the accelerating voltage pulse is applied. A further innovation with some QqTOF instruments is the use of multiple reflections in the oaTOF, which is part of a more general application of multiple reflection MS. Instead of collecting the ions after their first reflection, they are reflected back into the TOF for a second pass, each ion following a W-path rather than the V-path of the conventional oaTOF. This gives superior resolution of about 20,000 that starts to make it competitive with FT-ICR, though with an inevitable reduction in sensitivity compared with single-pass operation.

**FT-ICR**

Of the available *commercial* mass spectrometers, the FT-ICR is capable of the highest performance in terms of resolution and mass accuracy. It has been claimed that a mass measurement on a single peptide can give an accurate mass tag that is sufficient to identify a protein in a database, although there have been few demonstrated examples of this approach. For protein analysis, the performance specifications of most high-performance tandem mass *spectrometers* such as the QqTOF are relevant only to the analysis of fragments (peptides) but not to intact proteins. By contrast, the FT-ICR is capable of *isotopic* resolution of intact proteins. Ion activation can be induced by infrared laser radiation, resulting in *multiphoton* dissociation, or by pulsed admission of CID gases. However, it is difficult to introduce sufficient energy through collisions of large ions with light gases to achieve significant fragmentation and only the lowest energy pathways are populated, so conventional CID on large proteins is relatively uninformative. *Electron capture*-induced dissociation (ECD) was introduced to provide more comprehensive sequencing of small to mid-sized proteins, possibly with automated computer interpretation. This relied on capture of low-energy electrons by multiply charged ions, which then fragment to give c-type ions (i.e., *C-terminal* fragment ions formed by backbone cleavage between an amide nitrogen and the adjacent cit-carbon). Although initial studies employed thermal energy electrons, higher energy electrons (11 eV) produce radical cations such as z-type ions, further fragmentation of which gives w ions that, for example, can distinguish the isomeric amino acid residues of leucine and isoleucine. There are problems inherent in studies on larger ions, such as the difficulty of correctly assigning the *monoisotopic* mass due to the low abundance of this ion. Nevertheless, protein identification and substantial sequence information can be obtained in what have come to be known as "*top-down*" experiments on multiply charged intact protein ions, compared with the more commonly used "bottom-up" approach based on proteolysis and peptide analysis.

Although the FT-ICR spectrometer can itself be used for $MS^n$ experiments, ions are generally prepared externally and pulsed in via another ion transmission device. This may another mass analyser, such as a quadrupole, giving an extra dimension of tandem MS. Remote ionisation allows different types of ionisation techniques to be employed, but ESI is clearly the most popular. Although FT-ICR provides unique opportunities to *manipulate* ions in $MS^n$ experiments, the electronic pulse sequences are quite complex to set up and operate, and until recently it had not been established that the FT-ICR could be operated successfully in a routine analytical environment without highly experienced and skilled operators. Thus, it went against one trend in MS that seeks to make *sophisticated* instrumentation more readily accessible to nonspecialist operators, an increasingly important aim considering the *mushrooming* demand for MS compared with the small number of people being trained in this area. However, the Thermo Finnigan LTQ FTMS is a new hybrid instrument that combines a linear ion trap with an FT-ICR. As this comes from a company that has supplied large numbers of ion traps with software noteworthy for its ease of use, this instrument with a specification of 2 ppm mass accuracy with external *calibration* and 500,000 RP may do much to *popularise* FT-ICR.

**High-Energy MALDI Tandem Mass Spectrometry**

Despite a wide choice of instrument types and substantial differences in their performance characteristics, all the instruments referred to in this chapter give only low-energy CID. By contrast, the ion *kinetic* energies in conventional TOFs are in the *kiloelectron* volt range and should be ideal for high-energy CID. Furthermore, the MALDI process imparts more energy than ESI and may give ions that have sufficient internal energy to undergo unimolecular decomposition without further collisional activation. Ions that decompose by *unimolecular* processes may give rise to "*prompt*" fragments and metastable ions. For prompt fragments, unimolecular decomposition occurs before acceleration (i.e., within the 100 ns residence time in the DE source), and the fragment ions give peaks at the normal fragment m/z values. The metastable ions that dissociate later (i.e., after acceleration) will give fragment ions having the same velocity as the *precursor* ions and will arrive at the detector at the same time in a linear TOF but will be filtered out in a reflectron TOF. As a result of dissociation occurring during acceleration giving fragments with somewhat reduced velocities, some fragment ions are characterised by broader peaks occurring at higher m/z values than the expected fragment mass.

The sector oaTOF, referred to earlier in this chapter, gave high-resolution tandem mass spectra but with the sensitivity limit of 1 pmol. Until recently, post-source decay (PSD) with a reflectron TOF instrument was the only other widely available sequencing technique employing MALDI. This involves reducing the *reflectron* voltage in a series of steps to allow fragment ions formed after acceleration and, therefore, having reduced kinetic energies to be analysed; thus, it is a means of studying what were described earlier as metastable ions. For such experiments, the laser is operated at higher intensity than usual to increase the internal energy deposition and thereby increase the fraction of ions having sufficient energy to dissociate within the time available, typically a few *microseconds*. This rather *cumbersome* method that gives poor resolution and low mass accuracy is becoming superseded by new MALDI-TOF instruments, including a tandem TOF/TOF design that gives high-energy CID spectra.

A MALDI-TOF/TOF based on a conventional high-performance DEreflectron TOF has been described that can be operated in MS or MS/MS modes. This is illustrated in Figure elsewhere in this chapter. If ionisation is achieved with a high-frequency YAG laser operating at 200 Hz rather than the conventional nitrogen laser with a maximum frequency of about 20 Hz, *spectral* data can be acquired very rapidly. A typical YAG-laser wavelength is 354 nm rather than 337 nm, which has little effect on the spectra, although the laser power required for DHB matrix is greater because of low absorption at this wavelength. The resolution of the *precursor* ions in MS/MS mode (typically 0.5%) is defined by the opening and closing of a timed ion selector. The ions are then decelerated (e.g., to 1 keV) before entering an electrically floating collision chamber. The chamber contains a collision gas at a pressure designed to maximise the number of single collisions between ions and molecules but to minimise the number of ions experiencing

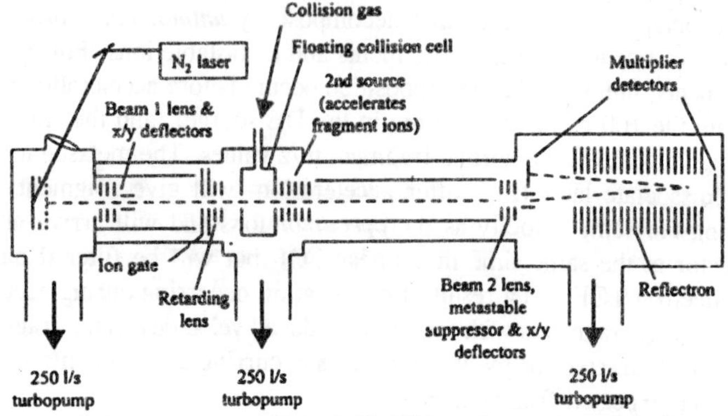


*Figure 1.13 : A time-of-flight (TOF)-TOF tandem mass spectrometer.*

multiple *collisions*. After traversing the *collision cell*, any remaining precursor ions and the ionic products of dissociation reactions traveling at the same velocity are accelerated by a high-voltage pulse into MS2, which is a reflectron TOF. Unlike PSD, this results in all fragment ions having almost the same high *kinetic* energy, within $\pm 0.5$ keV, allowing the whole mass range to be well focused by the reflectron. Consequently, the resolution of the fragment ion spectra is comparable to the separation of normal ions in a *typical* reflectron TOF. This instrument can be used with a range of *collision gases* and with different collision energies, although raising the collision energy results in a greater spread of final *kinetic energies* that may cause some loss of resolution in MS2.

For automated protein identification by the analysis of tryptic digests, the throughput of such an instrument is dramatically enhanced compared with manual operation of a more conventional tandem mass spectrometer. With a laser operating at 200 Hz and assuming 500 laser shots are sufficient to give a mass spectrum with good signal/noise ratios, an entire 96-spot plate can be scanned in 4 min. With real-time database searching, the majority of the proteins that have been digested to give multiple peptides can be identified by peptide mass fingerprinting during this time. The mass *spectrometer* can then be directed to carry out the further analysis of each spot by CID, first by making measurements on selected ions to confirm the previous assignments, and then to study peptides not assigned by mass mapping. The number of shots required per peptide and the total number of peptides to be studied will determine the time required for a complete analysis of all 96 spots, but it is likely to be completed within an hour or two. The provision of a multiplate cassette to hold 24 or more target plates enables unattended operation for as much as 24 h at a time. Thus, the comprehensive analysis of several thousand samples per day with a single instrument is becoming a realistic undertaking.

## ION DETECTION

In an instrument that produces and transmits a continuous ion beam, the ions arriving at a detector represent an electrical current. This current can be amplified and recorded (e.g., as a function of m/z in a conventional scan or at selected m/z values in a multiple ion monitoring experiment). The first stage of amplification is usually achieved with an electron multiplier having either discrete dynodes, as in Fig. 16, or a continuous dynode. The principle is based on emission of secondary electrons caused by an ion or an *electron* striking a metal surface. The multiplier illustrated is set up for positive ion operation as the first dynode is held at a negative potential attractive to positive ions. Successive dynodes are then progressively less negative, so the emitted electrons are accelerated between them. The number of secondary electrons emitted per event depends on the bombarding velocity at each dynode, which is controlled by the potential between the dynodes. Rather than the small number of *dynodes* illustrated here, in reality a discrete *dynode* multiplier may have 20 or more. If every impact releases an average of two electrons, the overall gain for a 20 dynode multiplier will be $2^{20}$ (i.e. $\sim 10^6$). A small increase in the voltage between the dynodes that increases the yield to an average of 2.25 electrons/impact will increase the gain to $2.25^{20}$ (i.e., $10^7$). The typical electron energy required to achieve this multiplication factor is about 100 eV, so a 20 dynode multiplier might be operated with an overall potential drop of 2 kV. Because there is a maximum output current that a multiplier can provide, the overall gain is adjusted to prevent this from being exceeded for the most abundant ions in the spectrum, so the output current is proportional to the ion current. In this mode the response time or bandwidth of the *electronic amplifier* is selected to give a continuous current that does not reflect the pulsed arrival of the ions at the *detector*. This conventional approach to ion detection and measurement in MS is best suited to ion currents approximately within the range $1 \times 10^{-12}$ to $1 \times 10^{-16}$ A, but it is generally unsuitable for very weak or intermittent ion currents or for experiments in which the m/z values are changing very rapidly, such as in a TOF mass spectrometer in which an entire spectrum may be obtained within about 100 zs. The alternative continuous *dynode* multiplier or channeltron made of glass is less expensive to produce and is generally smaller, so its use was pioneered in simpler and more compact instruments such as the quadrupole. A number of designs have been *explored*, but all involve a hollow tube, usually curved, coated on the inside with a high resistance material, down which the electrons travel. A voltage gradient down the tube causes *successive* sections to act like the discrete *dynodes* in Figure. The gain is dependent on the ratio of length to width. In general, the gain and lifetime of such a *multiplier* are likely to be less than those of the discrete dynode variety. It is also possible to have a 2D array of channels in a flat glass plate where the output *electrons* fall onto a phosphorescent screen and the signal is detected by a *photodiode* array. This is referred to as *multichannel* plate (mcp) detector. In an instrument that gives a spatial separation of the ions of different m/z, such as a magnetic sector, an mcp can be mounted at the *focal plane* of the *analyser*. The position at which an ion falls on the mcp will define its m/z, so the signal from such a detector can give a direct representation of the spectrum over a limited mass range. The gain of a single plate is lower than the gain of a conventional multiplier, but the plates can be stacked so that electrons emerging from the first plate can be multiplied again by a second plate. The spacing of the channels within the plate, which is typically 5–25 $\mu$m, defines the maximum resolution that can be achieved. The advantage of such an arrangement is that ions of a range of m/z values are detected simultaneously, increasing the sensitivity compared with scanning the ion current across a slit.

In most mass spectrometers, the multiplier/detector system acts as a *transient* recorder, the output current varying with the ion current. For experiments that produce lower ion currents, it is possible to monitor the arrival at the detector of individual ions, counting the number of ions arriving within a certain time window. Here, the *electron* multiplier is operated close to saturation, each ion falling on the *detector* giving an isolated pulse that can be detected and stored. For example, in the MDS-Sciex QSTAR instrument, the ions fall onto any one of four anodes of a four-*anode* multichannel plate detector with a time-to-digital converter capable of detecting single ions. Such a detector can record only one ion striking an individual anode within the counting period, but the presence of four *anodes* decreases the likelihood of two or more ions arriving at a single anode close enough in time to give only a single pulse. However, high ion currents that result in multiple ions striking a detector within the counting period will cause the stronger peaks to be truncated and distorted. The width of the detector pulse may also exceed the counting period, giving a dead-time after the arrival of an ion that can cause further signal distortion at high ion currents. The ion intensities are recorded as "*counts*" for ions arriving within each timing period, which in this instrument are 625-ps wide, and accumulated in "*bins*." The final mass spectrum is plotted as a histogram of the ion counts in each bin. Even at a resolution of about 10,000, each peak is several bins wide at low m/z values, the number of bins per peak increasing with mass as the square root of m/z. In normal MS mode the ion current is usually more than sufficient to determine the mass of each ion accurately, but it may be necessary to enhance the gain slightly for CID/MSMS experiments because the initial precursor ion current is likely to be distributed across many fragments. The threshold for ion detection is effectively adjusted by increasing the detector voltage for CID/MSMS, thereby increasing the average pulse height per ion. Thus, the "*counts*" recorded for a given peak do not necessarily constitute an absolute measure of the total number of ions of that mass transmitted and detected by the oaTOF.

MALDI-TOF is somewhat different as ion detection is intermediate between the analog techniques typical of quadrupoles and sector instruments and the pure pulse counting of oaTOFs. Relatively large numbers of ions are formed in isolated pulses in MALDI so the multiplier is not operated in saturated mode. Nevertheless, the response is nonlinear at higher ion currents as the multiplier is easily saturated by multiple ions arriving simultaneously. The output from the detector is approximately proportional to the number of ions arriving within a given time and is stored in a high-frequency digital oscilloscope. Thus, again the eventual output is a histogram of ion currents, which with a 500-MHz oscilloscope would be recorded every 2 $\mu$s. Ions of high m/z values such as singly charged protein ions moving relatively slowly through the mass analyser may not strike the first dynode of the multiplier with sufficient velocity to induce secondary electron emission. Thus, detection efficiency drops off at higher m/z values. The sensitivity is usually increased by the application of a conversion dynode held at a high potential of opposite polarity to the charge on the ions, toward which the ions are accelerated. This is referred to as post acceleration detection and is essential for negative ion operation, because the first dynode of the ***electron multiplier*** is usually negative and would repel negative ions.

Only FT ICR measures ion currents by a completely different method. At a fixed *magnetic field*, all ions of the same m/z value have the same cyclotron frequency and move together in a coherent ion packet. This cyclotron motion of ions is responsible for the induction of an electric current in *coils* surrounding the cell, m/z being determined from the cyclotron frequency. This is advantageous for weak currents because the signal can be *monitored* for longer times to improve the *signal/noise* ratio. Also the method is nondestructive, so the ions remain in the trap and are available for further experiments.

The properties of ESI and MALDI induced a quantum leap in biological MS, so that MS is now firmly established as an essential core technology in much *biological* research. When the last volume of Methods in Enzymology dealing with MS was published, ESI and MALDI were just emerging as the ionisation methods that would finally allow MS to *analyse* virtually any biomolecule without chemical derivatisation or transformation. Since then, the methods for ion separation and detection have been refined to more fully exploit the characteristics of these *ionisation* methods. Thus, the tools available for biological MS are the most *sophisticated* and *versatile* ever, even though new methods continue to emerge, such as selective binding and differential analysis directly from protein chips. In many respects, further improvements are now *incremental* and the current technology for *biological* MS is in a mature state, although absolute sensitivity still does and probably always will remain a challenge. Now MS is rapidly moving from a regimen in which samples are handled singly to the point at which *automated* systems are handling many thousands of samples per day. This is important for the application of MS to proteomics, which requires the identification of multiple proteins that are linked through their *cellular actions*, defining their levels of expression in different tissues in normal and abnormal states (e.g., during disease) and characterising their *posttranslational* modifications and their binding partners. Despite some valuable first steps in developing reliable *quantitative* methods in biological MS, this still presents substantial difficulties. Thus, looking ahead the immediate challenges are to *analyse* larger numbers of samples within shorter periods of time, achieve even higher sensitivity of detection, and improve the ability to quantitate the molecules under investigation. This will require enhanced data processing and *interpretation*, with validated scoring methods that will allow automation to the degree that reliable results will be generated by computerised methods with little or no *human* input.

# 2

# Analysis of Biomolecules

In recent years MS has emerged as a major analytical tool in biotechnology and *biochemistry*. Tandem MS (MS/MS) has become an especially valuable means for determining biomolecular structure. In this technique, a "*parent ion*" derived from the analyte of interest is selected in one *mass spectrometer* and then is broken up, usually by collisions with the ambient gas in a collision cell. The resulting "*daughter ions*" are examined in a second mass analyzer, yielding information about the structure of the parent. The triple *quadrupole* spectrometer ($Q_1q_2Q_3$) has been the most popular instrument for such MS/MS measurements. In this device, $Q_1$ and $Q_3$ are quadrupole mass filters, in which $Q_1$ serves to select the parent ion and $Q_3$ scans the daughter ion spectrum. The daughter ions are produced in the collision cell enclosed in a quadrupole ion guide $q_2$ (radio-frequency [RF] excitation only), which focuses the ions toward the axis.

A complete *mass spectrum* can be obtained only from a quadrupole mass filter by scanning; ion species in the spectrum are examined one at a time, discarding all others, which considerably reduces sensitivity (typically by a factor 1000) if a complete mass spectrum is needed. In the most common mode of operation of the triple quadrupole instrument, the first quadrupole $Q_1$ merely selects a given parent ion, so no scanning is involved. Indeed, a quadrupole is well suited to this role, because it couples efficiently to the collision cell. However, the entire mass spectrum of daughter ions is often of interest, and it must be obtained by scanning. The consequent reduction of sensitivity is a significant handicap when only a limited amount of sample is available.

This suggests that it is worthwhile to replace the final *quadrupole* $Q_3$ by a TOF spectrometer, in which the complete daughter ion spectrum can be measured in parallel (without *scanning*), yielding a large increase in sensitivity. The TOF spectrometer also provides this maximum sensitivity at full resolution, in contrast to the *quadrupole*, in which an increase in window width, the usual method of increasing sensitivity, degrades the resolution. Moreover, TOF mass analyzers have a number of other features that are particularly useful for the high efficiency analysis of *biomolecules*. First, their m/z range is effectively unlimited, apart from problems of ion production and detection. Second, a TOF instrument contains no narrow slits or similar restricting elements, which of course reduce sensitivity. Early TOF spectrometers suffered from poor resolution, but the use of *electrostatic* reflectors, and the *rediscovery* of the benefits of delayed extraction created substantial improvements in resolving power. Moreover, developments in fast electronics removed earlier limitations in the recording of TOF spectra, enabling the rapid response of the instrument to be exploited more fully. A TOF instrument, thus, provides in many cases an optimum combination of resolution, sensitivity, and fast response, particularly under conditions in which the whole mass *spectrum* is required.

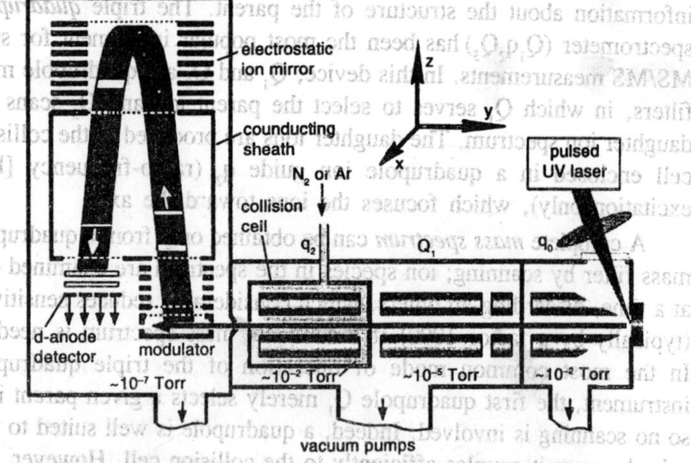


*Figure 2.1 : Schematic diagram of a quadrupole/time-of-flight (TOF) mass spectrometer, showing the collisional damping RF quadrupole $q_0$, the mass-selecting quadrupole $Q_1$, and the collision cell enclosed in the RF quadrupole $q_2$.*

For these reasons, hybrid quadrupole/TOF instruments have been developed as alternative devices for MS/MS measurements. In one such configuration (QqTOF), the first two sections (Qq) of a triple *quadrupole* spectrometer are coupled to a TOF spectrometer. In another configuration (QhTOF), an RF hexapole h encloses the collision cell instead of the RF quadrupole q. The functions of the first two sections of both the triple quadrupole and the *hybrid* instruments are the same—selection of the parent ion in Q and its dissociation in the collision cell within q or h. The difference appears in the final section, in which a TOF spectrometer replaces the final quadrupole mass filter. Figure elsewhere in this chapter shows a schematic diagram of a QqTOF instrument containing a single stage electrostatic mirror. In addition to the mass-selecting quadrupole $Q_1$ and the collision cell $q_2$, this instrument includes another quadrupole $q_0$ to provide collisional cooling of the ion beam. The mirror and the ion detector are similar to those used previously in single-MS measurements with pulsed ionization sources.

## ION PRODUCTION

As in other mass spectrometric applications for the study of biomolecules, the most effective ion production methods are ESI and MALDI. Using these techniques, intact gas phase *molecular ions* with molecular weights up to hundreds of *kiodaltons* can be formed.

Because the ion beams produced by ESI and MALDI have very different characteristics, it has been customary to examine them in different types of mass analyser, but the hybrid quadrupole/TOF instruments have the advantage that they can be adapted to use either type of ion source. We first discuss the type of TOF instrument necessary for efficient use of an ESI source, and then we discuss the problems and benefits of interfacing a MALDI source to this device.

### Coupling an ESI Source to a Quadrupole/TOF Instrument

Electrospray produces a continuous beam of ions. Like other continuous ion sources, it is most compatible with mass spectrometers that operate in a similarly continuous fashion, such as quadrupole mass filters. Indeed, the combination of an ESI source and a quadrupole mass filter has become a popular and satisfactory configuration, so there is no difficulty in coupling an electrospray source to the initial quadrupole sections of the hybrid quadrupole/TOF mass spectrometer. However, problems do arise in coupling the beam leaving the quadrupoles to the TOF section, because a TOF spectrometer requires a pulsed beam. The most straightforward coupling technique is to chop the continuous *electrosprayed* beam into short packets and to inject these along the spectrometer axis as in MALDI/TOF instruments. However, this procedure involves a tremendous loss in sensitivity, because gating ions into the TOF analyzer in this way to produce well separated ion packets imposes a serious reduction in the fraction of the total beam injected into the TOF analyzer. At best, this fraction then has the value ([length of packet]/[time between packets]). For example, Pinkston et ai. found that the need to extract short ion packets with a small energy spread meant that only 0.0025% of the total ion current contributed to the recorded spectra when they interfaced a continuous chemical ionization source to a TOF mass analyzer.

Orthogonal Injection. Fortunately TOF instruments can tolerate a relatively large spatial or velocity spread in a plane *perpendicular* to the *spectrometer axis*, as illustrated by the large sources (up to 1 cm diameter) typically used in fission fragment desorption (PDMS). This tolerance can be exploited by injecting electrospray ions into the TOF instrument perpendicular to the z axis (i.e., "*orthogonal injection*"). This *geometry* provides a high efficiency interface for transferring ions from a continuous beam to a pulsed mode. Another advantage is the small velocity spread in the z direction that is usually observed, making high resolution easier to obtain.

Such a configuration is illustrated in the QqTOF instrument shown in Figure elsewhere in this chapter. Here, the electric field direction in the single stage *electrostatic* ion mirror defines the *spectrometer axis* (the z axis). After leaving the collision cell, ions are re-accelerated along the —y axis to the desired energy. This energy is usually less than 10 eV, so the ions still have a relatively low velocity. They are then focused by ion optics into a parallel beam that enters continuously into a storage region in the ion modulator of the TOF analyzer, still perpendicular to the spectrometer axis. Initially, the storage region is field free, so ions continue to move in their original direction in the gap between a flat plate below and a grid above. When a voltage pulse is applied to the plate, a sausage-shaped ion packet is pushed out of the storage region into an accelerating column (the second stage of the modulator), in which the ions are accelerated to *kioelectron volt* energies. The longest flight time of interest determines the maximum injection frequency, typically 4–10 kHz. Thus, a series of ion packets is injected into the accelerating column at that frequency and, from there, into a field-free drift region of the TOF instrument, providing a maximum flight time of about 100 zs or more between pulses.

**Collisional Cooling**

Both transmission and resolution in the TOF analyzer in QqTOF instruments are very sensitive to the "*quality*" of the ion beam entering it (much more so than the final q in triple quadrupoles). Thus, the application of "*collisional cooling*" to TOF measurements has been an important development, because this technique can produce significant improvements in beam quality. Such cooling was introduced previously in *quadrupole* mass spectrometers, which have commonly used an RF beam guide to couple the ion source to the mass selecting quadrupole q. This beam guide consists of a set of quadrupole (or other multipole) rods to focus the ion beam onto the axis, and in early instruments, the beam guide was operated at relatively high vacuum, based on the argument that any gas in that region would cause beam attenuation. However, it was demonstrated by Douglas and French and Xu et al. that a higher pressure (0.01–1.00 torr) in the beam guide leads to *collisional* damping of the ion beam, thus improving its beam quality and consequently increasing the ion transmission. For similar reasons such a beam guide provides a useful interface to a TOF mass spectrometer, and *Krutchinsky* et al. have reported calculations and measurements that illustrate the collisional cooling obtained in that case.

In the collisional cooling ion guide, the RF field in the quadrupole focuses ions onto its axis while collisions with the molecules of the ambient gas reduce the ion velocities to near-thermal values. Thus, the beam leaving the ion guide has a greatly reduced spatial spread in the x-z plane, and much smaller velocity spreads in all three directions (x, y, and z), and these properties have no observable dependence on the original parameters of the beam delivered by the source. Indeed, collisional cooling decouples the mass measurement almost completely from processes occurring in the ion source, that is, the ions "*forget*" how they were formed.

The quadrupole/TOF instrument of Figure takes advantage of this technique by sampling ions from the ion source through the additional *quadrupole* $q_0$, in which the ions produced by the ion source (the parent ions) are cooled. In addition, radial and axial collisional damping of both parent and daughter ion motions occurs in the collision cell $q_2$, because this cell also is operated in the RF-only mode at a pressure of several millitorrs or more. In both cases, the cooling results in better transmission through the *quadrupoles*. Moreover, the ions retain their near-thermal energy spreads as they are reaccelerated to the desired energies after leaving the quadrupoles, so the benefits of cooling are retained for the ions entering the TOF analyzer.

## ORTHOGONAL INJECTION OF MALDI IONS

The rationale for trying to couple MALDI to a TOF instrument by orthogonal injection is not immediately obvious, because axial injection of MALDI ions into a TOF spectrometer has often been described as an ideal marriage between compatible techniques. In the usual geometry, ions are ejected from the target along the axis of the TOF instrument (normal to the sample surface) by bombardment with a pulsed laser. A TOF start signal is supplied by the laser pulse or by the necessary extraction pulse if delayed *extraction* is used. The rather long time between pulses (usually 100 ms) gives plenty of ûight time, even for very heavy ions. The initiation of delayed extraction has provided excellent resolution, and equally excellent mass accuracy ($\sim$ 10 ppm) can be obtained in favorable cases with sufficiently careful calibration.

By contrast, direct *orthogonal injection* of MALDI ions into a TOF spectrometer suffers from considerably greater difficulties than in the ESI case, because of the larger velocity and angular spreads of the MALDI ions, and because of the pulsed nature of the MALDI beam. Consequently, attempts to provide this coupling by direct injection of MALDI ions have usually yielded unimpressive results in single-MS measurements. Additional problems arise for MS/MS measurements in a hybrid instrument with a mass-selecting quadrupole, because the high-velocity ions from MALDI may not spend enough time in a short quadrupole to permit efficient mass discrimination.

Nevertheless, axial injection of MALDI ions does have its own problems. Some peak broadening is produced by the energy spread of the ions in the plume ejected from the target by the laser pulse, even with delayed extraction. Delayed extraction itself complicates the mass calibration and can be optimized for only part of the mass range at a time, so spectra may have to be recorded in segments, and high mass accuracy may be difficult to attain. Moreover, optimal focusing conditions, as well as the mass calibration, depend to some extent on laser fluence, the type of sample *matrix* and support, the sample preparation method, and even the location of the laser spot on the sample. Finally, perhaps the most serious handicap of the axial MALDI/TOF configuration is the difficulty in obtaining structural information by MS/MS measurements. Such measurements have normally been carried out by the so called post source decay technique in which a *parent ion* is selected by a gate in the flight tube and the products of its metastable decay are observed after reflection. The capabilities of this method are seriously restricted by the limited *resolution* obtained for parent ion selection (~200) and the limited mass accuracy of the daughter ion measurement (~0.2 Da), as well as the difficulties of controlling the fragmentation process and interpreting the results.

The possibility of *orthogonal injection* of MALDI ions was revived by the realization that conditions for it can be made much more favorable by collisional cooling. Thus, a MALDI ion source can be coupled directly to a TOF spectrometer through an RF quadrupole ion guide with collision-al cooling, similar to the one used for electrospray. For example, the MALDI source shown in Figure elsewhere in this chapter replaced the electrospray source originally installed in this hybrid instrument. When MALDI ions are injected into such a spectrometer, collisions in the ion guide damp the ion motion, improving the m/z range and the resolution. Measurements with this configuration give resolution and sensitivity comparable to axial MALDI performance and have several additional benefits:

1. The collisions in the ion guide also spread the ion pulse out along the *quadrupole axis*, producing a quasi continuous beam (as shown in Figure elsewhere in this chapter, which can then be treated just like an *electrosprayed* beam. Although it might seem odd to produce a continuous beam from the pulsed MALDI beam, only to pulse it again for injection into the TOF instrument, the final beam has highly favorable properties very different from those of the original one. In particular, the start time for the TOF measurement no longer needs to be correlated with the laser pulse, so the beam can be injected into the *spectrometer* at a high repetition rate (up to 10 kHz) as in ESI/TOF, although the laser itself may run at about 20 Hz or less. This reduces the number of ions in a single time measurement by about the same factor (~500 1/4 10,000/20) and, thus, eliminates problems of peak saturation and detector shadowing from intense *matrix signals*. As a result, the count rates obtained are compatible with digital pulse-counting methods (i.e., time-to-digital converters).

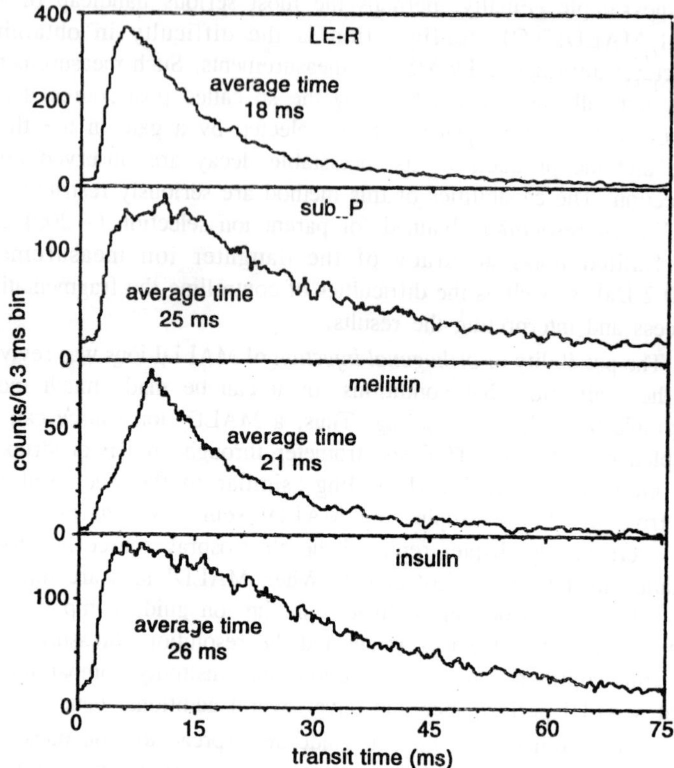


*Figure 1.2 : Measured transit times through a collisional cooling quadrupole. Note that for a 10-kHz injection, the time between injection pulses is 0.1 ms, so the beam entering the time-of-flight modulator is almost continuous.*

2. The residence time of ions in the guide is typically a few milliseconds or more, as shown in Figure elsewhere in this chapter, so the plume of ions has more time to expand and cool before it is injected into the flight tube, eliminating much of the *metastable decay* otherwise taking place in this region and, thus, simplifying the *spectrum*.

3. The performance of the instrument is independent of ion source conditions, as a result of the *decoupling* of the ion desorption event from the flight time measurement by the ion guide and the *collision cell*. Thus, there is much greater flexibility in the choice of different sample-preparation methods and different laser wavelengths, pulse widths, and fluences. Even *insulating* target supports can sometimes be tolerated.

4. The optimum conditions for orthogonal injection are independent of m/z, so the detailed adjustments required in axial MALDI are unnecessary. The detector is set for single-ion counting and the laser is simply set to any convenient fluence up to the maximum. The laser spot can be moved to a different spot on the target without any readjustment. The acquisition of MALDI spectra is, therefore, extremely simple, requiring very little operator expertise. The measured flight time has a simple quadratic dependence on mass; a simple two-point calibration gives optimum mass accuracy over the *entire mass* range. We have found that such an external calibration gives mass *accuracies* better than about 10 ppm in the instrument of Figure if performed within an hour of the measurement, provided that the room temperature is fairly stable.

5. Both ESI and MALDI sources can be used on the same instrument with minimal changes in configuration, because the ion beam entering the spectrometer has similar properties in both cases.

6. Perhaps most significant is the ability to carry out MS/MS measurements on MALDI ions in the *quadrupole*/TOF instruments just as on ESI ions, with similar high values of sensitivity (fmole) and control over conditions for collision-induced dissociation. Parent ions (up to a few thousand daltons) can be selected with unit mass resolution in the mass-selecting quadrupole; high resolving power (~10,000) and mass accuracy (~10 ppm or better) for the daughter ions are provided by the TOF analyzer. For structural determination, both MS measurements (mass mapping) and MS/MS measurements can be carried out interactively on the same sample and in the same instrument, without breaking vacuum.

On the other hand, an orthogonal injection instrument with collisional cooling is more complex than an *axial-injection* MALDI instrument, even though the quality of the data in the single-MS mode is rather similar. More seriously, there are losses associated with the *orthogonal injection* process that are not found with axial injection, as discussed later, so sensitivity is expected to be somewhat lower. Moreover, most existing orthogonal injection instruments use small accelerating voltages (~4 kV), so they have low efficiency for large singly charged ions (as produced by MALDI). The report of an orthogonal injection TOF spectrometer with 20 kV acceleration voltage is *encouraging*, but such a configuration is not yet available in a quadrupole/TOF instrument. In addition, there still may be difficulties in *transmitting* high m/z ions efficiently through the quadrupole ion guide, and problems related to *fragmentation* and adduct formation for large singly charged ions as a result of their long transit times.

## FEATURES OF THE ORTHOGONAL INJECTION/TOF CONFIGURATION

### Flight Tube High Voltage

In any TOF spectrometer, the need for DC acceleration of the ions means that either the flight tube or the previous stage of the instrument must be elevated to the DC acceleration potential, several kilovolts or more. In *axial* MALDI/TOF *spectrometers* (and in an early ESI/TOF instrument, the latter solution is adopted, that is, the previous stage (the source itself) is run at high voltage. However, this is not practical in quadrupole/TOF instruments because of the need to couple the TOF section to the quadrupoles, which normally operate close to ground potential. Thus, the former solution is chosen, raising the flight tube to the accelerating potential. In this case, the ions are accelerated to their final energy by the DC field produced by the potential Vz applied across the acceleration column, and positive ions are accelerated from ground to high negative voltages, rather than from high positive voltages to ground as in axial MALDI/TOF mass spectrometers. The subsequent motion is similar in both cases; ions of different m/z values are separated in time as they pass down the flight tube and are reflected by the mirror.

However, it is usually impractical to put high voltage on the whole vacuum chamber, so the flight tube must be built as a separate structure within the vacuum chamber (i.e., the "*conducting sheath*"). This structure is typically formed from a mesh because of the need for efficient evacuation of the interior, and particular care must be taken to avoid *penetration* of external *electric fields* through the mesh, which would degrade the performance of the instrument. The difficulty of keeping this rather large structure at high voltage is one reason that orthogonal injection TOF instruments have run at fairly modest accelerating voltages until the model mentioned earlier was developed.

### Direction of Entry and m/z Range

When the electric field in the mirror has the correct value, the total flight time t is $2L/v_z$, where L is the z component of the effective flight path. The distance that the ions move in the y direction during this time is $v_y t = 2L\, v_y/v_z$, so the ratio $v_y/v_z$ must have the value D/2L if the ions are to strike the detector, where D is the separation in the y direction between the modulator and the detector. Here, the axial velocity component vz is determined by the energy provided by the accel erating column ( $mv_z^2/2 = q V_z$), where $V_z$ is the voltage across the accelerating column, *m* is the ion mass, and q is the charge, that is, $v_z$ is proportional to $\sqrt{q/m}$. However, in general the initial value of the y component of *velocity* $v_y$ will have a different form, usually some fraction of the velocity of the supersonic jet in which the ions enter the vacuum system. Unless this is modi fied, the ratio $v_y/v_z$, will depend on mass and charge, resulting in a corresponding restriction in the m/z ratio observable at a given setting of $V_z$.

Again collisional cooling makes an important contribution. If the energy of an ion is reduced to near thermal values by collisional damping, its approximate energy mvy $^2/2$ on entering the storage region will be q Vy, where Vy is the voltage difference between the collision cell and the modulator. The ratio of velocities in the two orthogonal directions is then $v_y = v_z 1/= \sqrt{V_y/V_z}$, (independent of m and q), so the ratio of voltages can be adjusted to satisfy the required relation $(V_y/V_z) = (D/2L)^2$ for ions of all m/z values. Thus, all ions enter the field free drift region in a direction making a small angle with the z axis (set by the ratio of accelerating voltages), and in principle they all strike the detector. For example, collisional cooling has made it possible to observe ions over a mass range from about 1300 Da to more than 1 MDa at the same time (m/z range from 600 to more than 12,000) in an orthogonal injection TOF spectrometer, as shown in Figure elsewhere in this chapter. Note that this requires no additional deflection in the drift region, which could degrade the mass resolution.

**Time Focusing, Reso!ution, and Mass Accuracy**

Factors limiting the resolution of orthogonal injection TOF spectrometers have been discussed in detail by Dodonov etal. Even if the mechanical and electrical tolerances in the instrument are controlled precisely, the TOF resolution is still limited by the *initial spatial* and velocity spreads of the ions (*z* and $v_z$) in the storage region of the modulator, emphasizing the importance of *collisional* cooling before the ions enter this region. For example, a spread in the z coordinates of the ions (typically a few millimeters), leads to variations in energy (and consequently in velocity) of the ions leaving the modulator. The effect of these variations on the flight time can be reduced in a linear spectrometer by adjusting the fields in the modulator so that ions originating from different z positions in the storage region arrive at the plane of the detector at about the same time, thus improving the resolution; the ions exchange their original spatial spread for a velocity distribution on this plane. However, considerably better results can be obtained in a reflecting spectrometer. Here, the modulator fields can be adjusted to place the time focus plane fairly close to the source, thus minimizing the axial dimension of the ion packet. The plane then acts as an "*object plane*" for the mirror, and the electric field in the mirror can be adjusted to correct the ion velocity spread, as first shown by Mamyrin. This greatly increases the flight time and, therefore, the mass dispersion, without increasing the width of an individual peak. A single stage mirror, such as the one shown in Figure elsewhere in this chapter, corrects the velocity spread in the object plane to first order, which is good enough to give a resolving power of about 10,000 as long as the energy spread in the plane is less than about 5% of the total energy.

By using a tw o stage mirror, it is possible to adjust voltages to obtain second or higher order time focusing; spectrometer designs have achieved a resolving power between 15,000 and 20,000. Improvements in resolving power can also be obt ained by increasin g the length of the flight path (with a corresponding loss of sensitivity), in which case the main factor limiting the resolution is the residual energy spread in the beam. The flight path may be lengthened either directly or by introducing additional reflectors. For example, a resolving power of 20,000 has been specified for a commercial instrument that uses a second mirror to increase the flight path.

Nevertheless, the combination of collisional cooling with the single stage mirror and effective flight path approximately 2.8 m in the instrument of Figure elsewhere in this chapter yields a resolving power ($M/\Delta M_{FWHM}$) between 8000 and 10,000 for peptides with masses 1000–6000 Da, as shown in Figure elsewhere in this chapter for a MALDI UV source with a few nanoseconds pulse length. The same resolution is achieved for other *lasers* and *matrices*, as shown in Figure elsewhere in this chapter for the *molecular ion* of *insulin*, using a *glycerol matrix* and an infrared laser with a 200 ns pulse width. The highest resolution obtained with a similar orthogonal injection TOF geometry, but without collisional cooling (using an ESI source) was about 5000. Note, however, that the peak shape is determined mostly by the *isotopic* envelope for heavy ions produced by *electrospray ionisation* (m> 10,000 Da).

The approximate 10,000 resolution is sufficient to distinguish

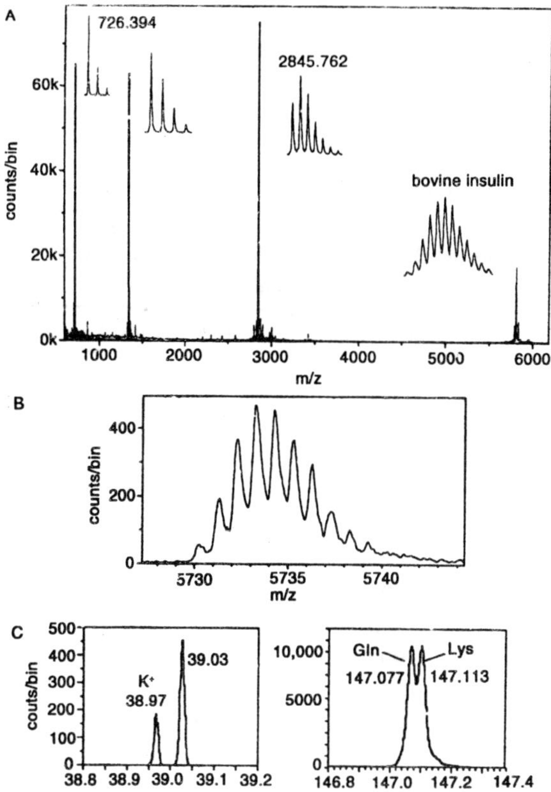


*Figure 1.3 : (A) Positive ion spectrum of a mixture of dalargin (726 Da), substance P (1348 Da), melittin (2846 Da), and bovine insulin (5733 Da). (B) Spectrum of bovine insulin with an infrared laser. (C) Small molecule mass spectra observed in an electrospray/time of flight instrument.*

separate *isotopic peaks* in ESI spectra for most *peptides* and, hence, determine their individual charge states. Such capability is particularly useful when interpreting complicated spectra from mixtures such as tryptic digests or *spectra* resulting from fragmentation of multiply charged ions. For smaller ions, the resolution decreases, but it remains sufficiently high to partially resolve two or more peaks within the same nominal mass, as shown in Figure elsewhere in this chapter. This is an advantage of TOF mass spectrometers over *quadrupoles*, where the resolution usually decreases linearly with decreasing mass in this range.

The high resolving power delivered by the TOF section of the QqTOF instrument is an important advantage of the quadrupole/TOF instruments over triple quadrupoles, in which high resolution is difficult to obtain along with adequate sensitivity. An even more important advantage is the high mass accuracy (better than 10 ppm or 10 mDa) that can be obtained, resulting from the combination of the high resolution and the decoupling of the ion production process

**Table 2.1 : Comparison of CID Fragmentation of the Singly-Charged Ion (m/z 1734.98) From the BMVW Peptide (Residues 24-41) With Mass Value Calculated Assuming $^{27}$Val Deletion.**

| *b ion* | *m/z (Obs.)* | *MH⁺ (Calc.)* | *Δ (mDa)* |
|---|---|---|---|
| b1 | — | 102.056 | — |
| b2 | 173.091 | 173.093 | -2 |
| b3 | 329.180 | 329.194 | -14 |
| b4 | 457.246 | 457.252 | -6 |
| b5 | — | 554.305 | — |
| b6 | 653.361 | 653.373 | -12 |
| b7 | 766.441 | 766.458 | -17 |
| b8 | 865.511 | 865.526 | -15 |
| b9 | 991.566 | 994.569 | 3 |
| b10 | — | 1091.621 | — |
| b11 | 1204.700 | 1204.705 | -5 |
| b12 | — | 1275.743 | — |
| b13 | 1346.764 | 1346.780 | -16 |
| b14 | — | 1403.801 | — |
| b15 | 1531.843 | 1531.860 | -17 |
| b16 | 1588.857 | 1588.881 | -24 |
| b17 | 1716.973 | 1716.976 | -3 |

from the mass measurement by collisional cooling, which makes the ion beam insensitive to changes in ion source conditions. For example, such accuracy has been crucial in identifying a mutation in the coat protein of a particular brome mosaic virus isolate. In that case, the measured mass of a particular tryptic fragment differed from the predicted value by 99.074 Da, most obviously achieved by deletion of valine 27 (99.068 Da) in the predicted sequence. However, the differences Δ between measured and calculated values for the b ions are all negative on this assumption, with an average Δ = -11 mDa, suggesting a search for other possibilities. Instead, replacement of

Arg 26 by Gly (genetic code change AGG to GGG) gives almost the same calculated mass change (-99.080 Da), but considerably better agreement for the b ions (Δ = -0.4 mDa), as shown in Table elsewhere in this chapter. This stimulated a more careful search of the MS/MS spectrum, which showed that there were very small peaks (initially not assigned) corresponding to $b_3$ and $y_{15}$ in the latter peptide, confirming the substitution of Arg by Gly.

**Table 2.2 : Comparison of CID Fragmentation of the Singly Charged Ion (M/z 1734.98) From the BMVW Peptide (Residues 24-42) with Mass Values Calculated Assuming A Mutation of R to G At Position 26.**

| *b ion* | *m/z (Obs.)* | *MH⁺ (Calc.)* | *Δ (mDa)* |
|---|---|---|---|
| b1 | — | 102.056 | — |
| b2 | 173.091 | 173.093 | -2 |
| b3 | 230.118 | 230.114 | +4 |
| b4 | 329.180 | 329.182 | -2 |
| b5 | 457.246 | 457.241 | +5 |
| b6 | — | 554.294 | — |
| b7 | 653.361 | 653.362 | -1 |
| b8 | 766.441 | 766.446 | -5 |
| b9 | 865.511 | 865.515 | -4 |
| b10 | 994.566 | 994.557 | +9 |
| b11 | — | 1091.610 | — |
| b12 | 1204.700 | 1204.694 | +6 |
| b13 | — | 1275.731 | — |
| b14 | 1346.764 | 1346.768 | -4 |
| b15 | — | 1403.790 | — |
| b16 | 1531.843 | 1531.849 | -6 |
| b17 | 1588.857 | 1588.870 | -13 |
| b18 | 1716.973 | 1716.965 | +8 |

**Duty Cycle and m/z Discrimination**

As remarked earlier, all ions are recorded in parallel in a TOF analyser, so the overall efficiency is much larger than in scanning instruments. This advantage is possessed by both *orthogonal injection* TOF *instruments* and *axial* TOF spectrometers. However, the axial spectrometers are usually more efficient, because all of the output of their pulsed ion sources can be injected into the flight tube.

On the other hand, *orthogonal injection* normally involves losses and some m/z dependence associated with the duty cycle, defined here as the fraction of the beam entering the TOF spectrometer that is available for acceleration into the flight tube. To avoid spectral overlap, an injection pulse cannot be applied until the slowest ion from the previous pulse has reached the detector. For instruments without deflection, the velocity component vy of an ion in the flight tube is the same as its velocity in the *storage region*. Thus, the slowest ions (the ions of highest mass) entering the storage region would cover the distance between the modulator and the detector (D) by the time the next injection pulse arrives, if the storage region were long enough. However, only a finite slice of this beam of length $\Delta l$ can be accelerated and detected, where $\Delta l$ is determined by the size of the *apertures* in the detector and/or the *modulator*. This sets an upper limit on the duty cycle for this m/z value as $\Delta l/D$.

If the incoming beam is monoenergetic, vy is proportional to $1/\sqrt{m/z}$, so lighter ions will cover a distance larger than $D$ between TOF pulses. Thus, the duty cycle is proportional to $\sqrt{m/z}$ and there is *systematic discrimination* against low-m/z ions:

$$\text{DutyCycle}(m/z) = \frac{\Delta l}{D}\sqrt{\frac{m/z}{(m/z)_{\max}}} \tag{1}$$

where $(m/z)_{max}$ corresponds to the ion in the spectrum with the largest m/z value to be measured. Nevertheless, this discrimination is *predictable*, and signals at low m/z values are normally stronger than at high m/z values, so in most cases this defect is preferable to the decrease in transmission at high m/z that is characteristic of *quadrupole* mass filters.

In most reflecting TOF instruments with *orthogonal injection*, the *duty cycle* is between 5% and 30%, depending on the m/z value of the ion and the instrumental *parameters*. Whitehouse et al. demonstrated an improvement in the duty cycle by *trapping ions* in a two-dimensional ion trap (an RF octopole with *electrostatic* apertures at each end), and then gating them in short bursts into the TOF spectrometer. These results have been extended by Chernushevich (2001); by varying trapping and gating parameters, different modes of operation were achievable, from 100% duty cycle over a narrow m/z range to <50% duty cycle over a wider mass range, although ions outside of these ranges were lost partially or completely. In spite of the reduced m/z range, this *technique* can be very useful when applied in certain MS/MS modes of operation, where only selected fragment ions have to be *monitored*, as discussed later in this chapter.

**Ion Detection**

The TOF detector normally consists of two large area microchannel plates in a *chevron configuration*. Signals may be recorded by either digital or analog techniques, each of which has advantages, depending mainly on the maximum intensity of the ion signal. In the digital technique used in the instrument of Figure, the detector signal passes to a preamplifier and a constant-fraction discriminator and is then registered in single ion counting mode by a time-to-digital converter (TDC). This has the advantage that the time measurement is *independent* of the *detector pulse* width, so this parameter does not limit the resolution.

The TDC can record hundreds of stops for each injection pulse, but only one stop per injection pulse for a given m/z species (i.e., within the "dead time" of the TDC, about a few nanoseconds). To avoid *spectral distortion*, the count rate for an *individual ion species* has an effective limit somewhat less than the frequency of the injection pulse. For axial MALDI, this frequency is the laser repetition rate, usually less than 20 Hz, so TDCs are not very useful for such measurements. In the configuration of Figure elsewhere in this chapter, on the other hand, collisional cooling spreads the ions from a single laser pulse over tens of *milliseconds*, so typical injection frequencies of 4–10 kHz can be used. Thus, the limitation is much less severe than in axial MALDI, and in most cases, spectral distortion is negligible, even for the highest laser fluence obtainable from the usual 20 Hz laser. However, for higher *average* power lasers, and especially for simple spectra, where the total ion current is concentrated in a few peaks, distortion will be observed, and it limits ultimate throughput for orthogonal MALDI.

The problem is more severe for *electrospray* because the average ion currents can be considerably higher and are more difficult to control. The problem is particularly acute in LC/MS experiments, where the spectra are simple, and the maximum count rate determines the available *dynamic range*.

The most *straightforward* solution to the *distortion problem*, when using a TDC, is to divide the anode into a number (n) of sections, each connected to an independent TDC channel. Assuming a uniform distribution of ions over the *anode*, this gives an n-fold increase in the allowed data rate. For example, the instrument of Figure elsewhere in this chapter employs a four-segment anode and a four-channel TDC (both from Ionwerks, Inc., Houston, Texas). In principle, n can be increased to any value needed, and up to 256 segments have been reported, but cross-talk becomes a serious problem beyond about four or eight channels, and the expense of the TDCs soon becomes significant. An alternative solution is to use anode segments of different sizes. In this case, only the small *anode* is used for high count rates, and the intensity is scaled up according to the *geometry*.

For count rates that are not significantly greater than one ion per injection pulse per species per anode segment, it is possible to correct for the distortion using simple software *algorithms*. Other *strategies* are also possible, such as measuring weaker peaks (e.g., a $^{13}C$ isotope peak) to scale intense peaks or introducing *dynamic* current limiting procedures.

Ions may also be recorded by an *analog* device such as an integrating transient recorder, which is better suited to high-intensity signals, and can run at the maximum repetition rate (*reciprocal* of the maximum flight time). However, such devices are limited in resolution by the width of the *detector* pulse, and they require much more careful matching of the detector and recorder gains, which depend on signal intensity. Moreover, because of inherent noise in the *analog-todigital* conversion, they are poorly suited to low count rates, such as the measurement of pulses caused by single ions striking the detector. Note that existing transient recorders are well suited for conventional (axial injection) MALDI because that technique typically operates at a much lower repetition rate (a few hertz), and high instantaneous current, whereas a high repetition rate in the range of several kilohertz is critical for obtaining a reasonable duty cycle in TOF instruments with *orthogonal injection* of continuous beams.

**Sensitivity**

The sensitivity of the QqTOF geometry for *electrospray ions* has been considered already. In single - MS mode, or for daughter ion *scanning*, the parallel detection of the TOF system gives a 10–100-fold improvement in sensitivity over a triple quadrupole instrument, inspite of the duty-cycle losses associated with orthogonal TOF. For parent -ion scans, using appropriate *trapping strategies*, the QqTOF sensitivity approaches that of the **triple quad** rupole, as discussed later.

The situation is somewhat different when comparing orthogonal MALDI with conventional *axial -injection* MALDI. In axial injection MALDI, essentially all desorbed ions are injected directly into the

TOF analyzer. In orthogonal MALDI, there are intrinsic duty cycle losses, as well as losses associated with the transmission of the quadrupole ion guides. Moreover, the TOF analyzer in orthogonal TOF uses at least one *additional grid*, and the voltage is typically lower, although this is not a disadvantage in principle.

In spite of these *intrinsic* losses associated with orthogonal MALDI, there are some features that provide compensation. The single -ion counting method is more sensitive than *analog* detection, particularly when weak signals a represent together with strong signals. In a ddition, because the quality of the spectrum in orthogonal MALDI is independent of the target conditions, it is possible to make more efficient use of the deposited sample by increasing the laser fluence and irradiating the sample until it is essentially all removed from the target.

Thus, in practice, the sensitivity of orthogonal MALDI approaches that of axial MALDI. Using the simplest type of dried droplet sample preparation with 0.5 $\mu$l sample deposited on the target, about 70 amole substance P yields a usable mass spectrum. Significantly cleaner spectra can be achieved by using smaller spots deposited from smaller volumes. In experiments with Karger et al. to test an off-line interface for CE MALDI, we deposited a mixture of four angiotensin peptides as a 100-$\mu$m wide strip. A single portion of this strip corresponding to the width of the laser beam ($\sim$ 100 $\mu$m) contained about 80 amoles of each of the peptides. A spectrum obtained from such a spot using 40 laser shots and with one laser shot. These spectra still provide excellent signal/noise ratios, suggesting the lower limit of sensitivity is well below 80 attomoles.

**Throughput**

The QqTOF mass spectrometer does not itself determine the throughput in most electrospray experiments, because the throughput is usually limited by the flow rate of the ion source. There has been some development of multi spraying technologies, but these have not yet gone beyond the existing capability of the QqTOF instrument.

The situation is different in orthogonal MALDI, because there is usually no direct coupling between the target preparation and the MS measurement. If separation is carried out off line, as in the example mentioned earlier, the time scale of the separation is independent of the time scale of the MS measurement, and multiple separations can be carried out at the same time. Under such conditions, particularly in the high volume applications that have become the focus of several *proteomics* projects, the mass spectrometer itself may well be the *bottleneck*.

As mentioned earlier, the ultimate limit in throughput for a QqTOF instrument with single ion counting is presented by the data system. For a repetition rate of 10 kHz and a 4 anode detector, the maximum count rate for a single ion species is about 40,000 ions/s, making use of software corrections for spectral distortion. Then for a dynamic range of 1000, assuming 40 ions is sufficient to define a peak (for examples in which this is more than adequate), spectra can be acquired at the rate of 1/s. For a dynamic range of 100, the rate is 10 spectra/s.

In principle, these rates can be reached in any experiment by simply increasing the repetition rate of the *laser*, assuming sufficient sample is available to sustain the count rate for the duration of the acquisition. In practice, for single-MS experiments, the limit is often reached using a simple nitrogen laser with 150 $\mu$J pulses up to 20 Hz. Indeed, even for weak signals such as those shown in Fig. 5, a spectrum like the one in Figure elsewhere in this chapter can be acquired in 2 s, and the one in Figure elsewhere in this chapter at the rate of 20/s. For these data, the extraction pulse rate was 4 kHz, so the highest count rate was about 10% of the maximum.

However, optimum acquisition of MS/MS spectra requires much more initial ion signal, because only one parent ion is selected, and the daughter ions are divided over all the peaks in the final spectrum. As a result, a 20-Hz *nitrogen laser* will not normally reach the limiting count rate, and good-quality MS/MS spectra typically take a minute or more to obtain using this laser. Again, the time required can be reduced to the aforementioned limits by using higher repetition rate lasers. Reasonably *inexpensive* kHz Nd-YAG lasers are now available, although the energy per pulse is only in the tens of *microjoules* range, so the laser beam must be focused to a smaller spot on the target. We have demonstrated that such a laser, appropriately focused, brings the required time to acquire good MS/MS spectra in the instrument of Figure elsewhere in this chapter into the range of a few seconds. At such high repetition rates, the sample is consumed rather quickly, so sample rastering becomes necessary.

## QUADRUPOLE OPERATION

When used as a mass filter, quadrupole rods have both RF- and DC-voltage components applied, and the reduced Mathieu *parameters*

$q_M$ and $a_M$ characterize the amplitudes of both components:

$$q_M \cong \frac{4eV}{(m/z)\omega^2 r_0^2} \quad a_M \cong \frac{8eU}{(m/z)\omega^2 r_0^2} \tag{2}$$

where *e* is the charge of the electron, V and ω are the amplitude and angular frequency of the RF voltage, U is the value of the DC voltage, and $r_0$ is the inscribed radius of the quadrupole. When U = 0, the quadrupole acts simply as an ion guide.

**Single MS Mode**

As just mentioned, the simplest mode of operation of the QqTOF instrument (single MS or TOF MS) occurs when the mass filter Q is operated in the RF only mode (U = 0, $a_M$ = 0), so the quadrupole sections simply act as transmission elements that deliver ions to the TOF spectrometer. This mode of operation is applicable, for example, to the measurement of the *proteolytic* fragments produced from *enzymatic* digestion of a protein (*peptide* mass mapping). The resulting spectra benefit from the high resolution and mass accuracy of the TOF instrument, as well as its ability to record all ions in *parallel* without scanning, as pointed out earlier. Such single MS spectra can be acquired either with or without a collision gas in q. In the former case, the collision energy is kept below 10 eV to avoid fragmentation, and both sensitivity and resolution benefit from the additional *collisional* damping provided by the gas in the collision cell.

In this RF only operational mode ($a_M$ = 0), quadrupoles serve as high mass filters: Ions are rejected if they have m/z values below a cutoff value corresponding to $q_M$ = 0.908. Although there is no such sharp cutoff for ions with high m/z values, their transmission suffers because of poorer focusing, because the depth of the effective potential well is inversely proportional to m/z. Ions can be lost in any one of the three multipoles on their way to the TOF section. The resulting effective bandpass within which ions are transmitted simultaneously into the TOF instrument normally covers approximately an order of magnitude in m/z. If a wider m/z range is required, the RF voltage can be switched between two or more RF levels during spectrum acquisition, but in this case, light ions are lost when the RF voltage is "*high*," and heavy ions are poorly focused in a shallow potential well when the RF voltage is "*low*."

**Tandem MS Mode**

For tandem MS in the instrument shown in Figure elsewhere in this chapter, Q is operated in the mass filter mode, so ions are transmitted only within a narrow m/z window, corresponding to $q_M = 0.706$ and $a_M = 0.237$. A *monoisotopic* parent ion of interest may be selected or the window may be widened to increase the sensitivity by transmitting the corresponding full *isotopic cluster*. Note that a measure of the distribution of isotopes selected by Q is provided by the distribution actually observed for the undissociated ions in the MS/MS spectrum.

After selection in Q, the parent ion is accelerated to an energy between 20 and 200 eV before it enters the collision cell q, where it undergoes collision induced dissociation during the first few collisions with neutral gas molecules (usually *argon* or *nitrogen*). The resulting daughter ions (as well as the remaining paren t ions) are collisional ly cooled and focuse d as they pass throu gh the rest of the collision cell, thus im proving the qua lity of the ion beam entering the TOF spectro meter.

The maximum m/ z trans mitted through a quadrupole mass filter with reasonable efficiency de pends on the choice of parameters in Eq. (1). In most commercial triple quad rupole or qua drupole/TOF spectrometers, these parameters have been chosen to give a maximum m/z value of 4000 or less. Thus, tandem measurements can be carried out in these instruments on ions of very high mass produced by electrospray (in correspondingly large charge states), but only up to masses of a few thousand *daltons* on the singly charged ions produced predominantly by MALDI. This mass range is usually adequate for sequencing applications, though in several instruments, the quadrupoles have been modified to give a larger m/z range for examination of ions with higher values of m/z, and the most recent quadrupole/TOF commercial instruments also have an extended m/z range as a result of the need to analyze MALDI ions.

In general, the problems of tuning the quad rupoles in the quad rupole/ TOF inst ruments are similar to those encountered with triple quadrupoles, as described in some detail by Yost and Boyd. How ever, the higher pressures normally used for collisional cooling lead to long residence times in the collision cell, which tends to produce adducts and make it difficult to switch rapidly from one parent ion to another, as required in some types of measurement. Consequently, a weak axial electric field in the collision cell has been found to be beneficial, and various methods of producing such a field have been implemented. The presence of the axial field means also that optimal collisional cooling can be attained for ions of all m/z values with the same buffer gas pressure. However, it is important that the method of producing the axial field does not restrict the m/z range transmitted, because the quadrupole/TOF instrument detects ions of all m/z values in the same measurement. The most recent design appears to provide a satisfactory solution to this problem.

**Parent Ion Scans in the Quadrupole**

The most common mode of measurement in triple quadrupole mass spectrometers is the daughter (or product) ion scan that measures the daughter ion spectrum that results from collision induced dissociation of a selected parent ion. As discussed earlier, quadrupole/TOF hybrid instruments have considerable advantages over triple quadrupoles for this type of measurement.

However, in sequencing unknown samples, it is helpful to be able to identify relevant parent ion peaks that may be obscured by chemical noise. For this purpose, an alternative mode of triple quadrupole operation can be used. This is the parent (or *precursor*) ion scan, in which the final quadrupole setting is fixed and the first quadrupole is scanned. This measures the spectrum of parent ions, giving rise to selected daughters, such as the *immonium ions* that characterize particular amino acid residues or *phosphotyrosine*. Such parent ion scans are useful tools for identifying components of a mixture that lose a particular *diagnostic* fragment. Here, the simple arguments for preferring a TOF spectrometer as a final stage in a composite instrument do not apply, because only a single daughter is being monitored, and a similar scanning function is carried out in the first quadrupole in both the triple quadrupole and the quadrupole/TOF hybrid instrument. Moreover, the efficiency for measuring the products of collisional induced dissociation is between one and two orders of magnitude lower in the quadrupole/TOF hybrid than in the triple quadrupole, because of the reduced duty cycle in the former instrument, as discussed earlier, and because of losses at grids and in the detector. Thus, there are advantages in doing parent ion scans in triple quadrupoles.

Nevertheless, the duty cycle losses in the quadrupole/TOF instrument can be largely removed by trapping ions in the collision cell, as described earlier. This reduces the m/z range, but in this case the range often needs only to encompass the desired daughter ion, so there is no disadvantage to *trapping*. The higher resolution obtainable in the TOF spectrometer is also helpful in discriminating between the diagnostic daughter ions and background. For example, there are 21 possible ions from unmodified peptides that are within 0.1 Da of the *phosphotyrosine immonium* ion at 216.043 Da. The final quadrupole in a triple quadrupole instrument will not likely resolve them, resulting in false positive identifications of phosphotyrosine, but the TOF instrument easily discriminates between the possibilities.

Another report describes a second important application of trapping in the collision cell—''*peptide end sequencing*''. If MALDI mass mapping fails to identify a protein unambiguously after a proteolytic digestion, an MS/MS measurement is carried out in which the intensity of low mass daughters (e.g., 120–500 m/z) is accentuated by trapping. Typically the N terminal $b_2$ and $b_3$ ions, as well as the $y_1$ and $y_2$ C terminal daughters, can be observed, and such end sequencing of one or two tryptic peptides is usually sufficient to provide unique identification of any protein included in the database.

In addition, the TOF instrument can check for the presence of multiple diagnostic fragments at the same time, provided they are within the m/z range trapped. Thus, quadrupole/TOF instruments are becoming increasingly useful for parent ion scans.

# 3

# INDUSTRIAL PERSPECTIVES ON ASSAYS

Although *microbial cells* and *enzymes* have been used by humans for millennia to produce a variety of foodstuffs, it was not until the twentieth century that *biocatalysis* was introduced in the chemical industry. Since that time it has grown *biocatalytic* processes currently successfully operated at industrial scale are the to become an established technology for *chemical* manufacturing. Examples of production of the lowcalorie *sweetener aspartame* catalyzed by the protease *thermolysin*, the production of the bulk chemical *acrylamide* using a nitrile *hydratase*, and the production of D-*p*-*hydroxyphenylglycine* using microbial cells containing a *hydantoinase* and a *carbamoylase*.

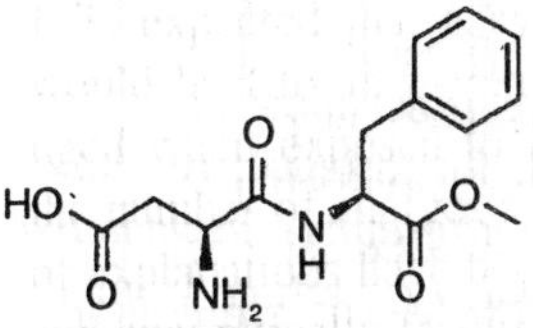

1 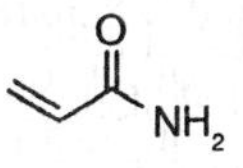 2 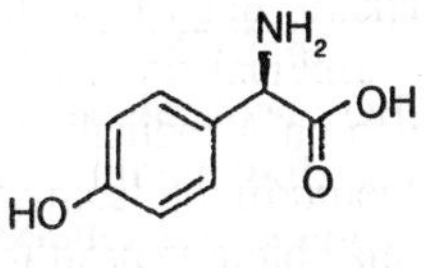 3

The recent need for more sustainable processes has driven the implementation of *biocatalysis* in industry further, in many cases by replacing existing chemical processes. An example of such secondgeneration "*green processes*" is the *chemoenzymatic* synthesis of the semisynthetic β-*lactam antibiotic cephalexin*. In this new process concept the sidechain *dphenylglycine*, activated as amide or ester, is coupled to the β-lactam nucleus 7-amino *desacetoxycephalosporanic acid* (7ADCA) by a *penicillin* G *acylase*. This process fully exploits the advantages of a *biocatalyst* in an industrial setting. The coupling reaction of activated sidechain and nucleus is done in water at ambient temperature, thereby eliminating the use of *halogenated* solvents and

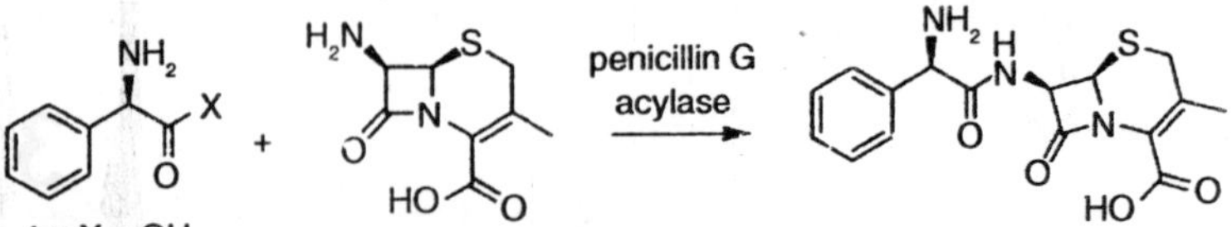

*Figure 3.1 : Penicillin G acylasecatalyzed synthesis of the semisynthetic β-lactam antibiotic cephalexin.*

low temperatures. The strict selectivity of the biocatalyst contributes to the high efficiency and increased sustainability of this *chemoenzymatic* process making protection of the two building blocks and deprotection of the coupling product superfluous. Altogether, over 140 processes using enzymes in isolated form or *microbial* cells have been commercialized up to now, most of which deliver *smallscale chiral* products for applications in the *pharmaceutical* sector.

Chemical custom manufacturing (CCM) is the industry segment in which the fine chemicals industry produces a compound on demand for a client (usually a *pharmaceutical* company) in a onetoone relation (as opposed to *multiclient* products). In this highly *dynamic business area*, speed and responsiveness are among the most important key success factors. General timelines for CCM are given in Table elsewhere in this chapter. It can be seen that proof of principle including a sample of ten needs to be delivered to the customer within just three months.

In CCM the use of *biocatalysis* for the production of (*chiral*) fine chemicals is still limited compared with, for instance, resolution via diastereomer crystallization or *chromatography*, *asymmetric organometal catalysis*, or conventional *chemistry* starting from the chiral pool. The main reason for this difference is the long lead time to find and implement a biocatalytic process compared with a chemical process, despite the fact that biocatalysts in general show excellent *enantio*, *regio* and *stereoselectivity*, which, in principle, often enables the design of shorter and more costefficient processes.

In order to be able to implement a biocatalytic reaction step successfully in such a product, extremely strict timelines to (bio)catalyst identification and subsequent process development need to be met.

Although the three months' time line and approach schematized in Table elsewhere in this chapter is an average scenario and deviations frequently occur, it clearly demonstrates the challenge for (bio)catalysis in CCM: the time available for (bio)catalyst identification and (smallscale) process development is extremely short.

**Table 3.1 : Typical timeline for process development within chemical custom manufacturing.**

| Time | Status | Remark |
|---|---|---|
| 0 | Pharmaceutical customer request | Chemical route exists but usually high costs<br>Quest for alternatives begins |
| 2 weeks | Selection of 2–3 routes | Based on rough cost price estimates<br>Plan and quote to customer<br>After customer's positive response, experimental work begins |
| 2 months | Feasibility demonstrated | Final choice made<br>More accurate cost price calculated |
| 3 months | Lab sample delivered for approval by customer | Based on specs or use tests |

In this chapter we will deal with the development of enzymatic assays and *activitybased* screenings from an industrial perspective. We will focus solely on CCM, and try to describe how this specialized branch of the chemical industry must cope with the demanding timelines for *biocatalyst* identification. In the first part, a number of prerequisites for the ideal screening will be discussed, including one of the most important ones: integral screening time. In the second part of this chapter, these general principles will be exemplified by a number of *assay/screening* methods developed for DSM's chemical custom manufacturing business unit DSM Pharma Chemicals (DPC).

## PREREQUISITES FOR AN EFFECTIVE BIOCATALYST SCREENING IN CHEMICAL CUSTOM MANUFACTURING

Identification of a biocatalyst for a certain chemical conversion requires input of sufficiently highquality *biodiversity* to start from. In the last decade *tremendous* progress has been made in the generation of such biodiversity. Construction of expression libraries directly from environmental DNA is applied in many *laboratories* as a source of diversity. Use of this socalled metagenome now enables identification of interesting *biocatalysts* from *microorganisms* that could not be cultivated in laboratories before, increasing the biodiversity dramatically. The rapidly increasing quantity of genome sequence information, together with improved software tools to predict the function of the encoded proteins, offers another source for new enzymes. This *genome* sequence information can be transferred into ready to screen *enzymatic* platforms.

Directed evolution is another relatively recent approach to the creation of *biodiversity*. Numerous formats ranging from simple errorprone *polymerase* chain reaction (epPCR) of a single gene to *shuffling* of large families of *genes* and even whole *genomes* have been published in the last 10 years to access the desired sequence space. These molecular biology tools have become quite robust, which makes the generation of sufficiently highquality biodiversity no longer a limiting factor in many biocatalyst identification programs. This is also true in CCM if the inputs for the screening (for example expression libraries and *enzymatic platforms*) have been *generated proactively* to meet the short timelines in this industrial segment.

Screening for *biocatalysts* has also attracted considerable attention from academia and industry in the last decade, undeniably resulting in significant progress in this field. Nevertheless, the development and implementation of a new screening within the limited time available in CCM is still a significant challenge. The following criteria are pivotal for successful screening in this demanding field.

**Screening for Real Conversion**

One of the most commonly used statements in screening is "You get what you screen for". In the open and patent literature, numerous artificial substrates have been described for *colorimetric* or *fluorometric* detection of a wide range of *enzymatic* activities, but use of this type of substrate bears the inherent risk that a biocatalyst identified does not catalyse the desired reaction. This can be caused by steric incompatibility (artificial substrates are often larger than the actual ones) or *catalytic* incompatibility (certain types of *artificial substrates* are more active than the actual substrates). Although improved artificial substrates have recently been reported, *screening* for the real conversions still results in the most reliable hits. Furthermore, synthesis of the artificial substrate can be difficult and time consuming, which will delay the outcome of the screening, thereby making it even harder to identify the desired *biocatalyst* on time.

**Robust and Reliable**

The ideal screening is robust and reliable. It must not suffer from matrix effects, such as small differences in pH, salt concentration, and residual growth medium. Furthermore, interexperimental differe-nces should be small. We noticed that the *robustness* of screening protocols sharply decreases with increasing numbers of handling (e.g. *centrifugation*, *pipetting*, and *filtration*). Therefore, simple screening protocols are preferred whenever possible. Altogether, the screening method must clearly discriminate true positive clones without delivering a large number of false positives.

**Quantitative**

To be able to *discriminate* positive *clones* the applied screening protocol must result in a clear change of measured response. However, different types of *projects* require different levels of accuracy. Screening a (*metagenomic*) library for new biocatalysts, for example, places less strict requirements on accuracy than screening in a directed evolution *program* to improve the *catalytic* performance of an enzyme. In the first example, the *scenario* is more or less black and white the few positive *clones* that are expected should give a signal that is sufficiently different from the background signal. The exact intensity of the signal of the positive *clones* is not very useful, however, since clones containing the same gene will often have a different genetic constitution that can *severely* influence the expression of the genes on the insert fragment. Therefore, an assay that delivers *semiquantitative* results is sufficient for this type of screening. In the case of a directed evolution project for improved enzymes the situation is different. In this case, the enzymatic reaction is tuned so that the conversion with the *wildtype* clones amounts to about 10% conversion. Consequently, these wildtype clones will already give a (low) signal, requiring more quantitative screening methods. The fact that all clones to be screened contain the (*mutated*) gene(s) of interest in an identical *genetic* constitution is another reason why quantitative screenings are advantag-eous in this case. As the development of highly quantitative screenings can be time consuming, the accuracy aimed for should be tuned to the needs of the specific application.

**Sensitive**

As the most reliable results are obtained by *mimicking* industrial conditions in the screening reaction as far as possible, substrate concentrations are usually high (> 50 mM). Consequently, the sensitivity of the *analytical* method is typically not an issue. Where the substrates or products are poorly soluble in water the situation is different, and a sensitive *analytical* method is required. Using mass spectrometry (MS) positive clones can be identified starting from substrate concentrations as low as 0.5 mM. Addition of a cosolvent enabling higher substrate concentrations is another possibility, assuming that it is not *interfering* with the screening.

**Integral Screening Time**

This is the most important factor for screenings in CCM. It can be defined as the time needed from the start of the screening development to the generation of the first screening results. As will have become clear from the introductory section, only screening methods with a short integral screening time are compliant with CCM timelines. At first *glance*, *ultrahigh* throughput screening (UHTS, in this chapter defined as a screening with a throughput of > $10^4$ samples a day) formats look suitable for application within CCM as they are characterized by an extremely short analysis time per sample. However, this is certainly not true in all cases, especially when the development of UHTS methods is time consuming.

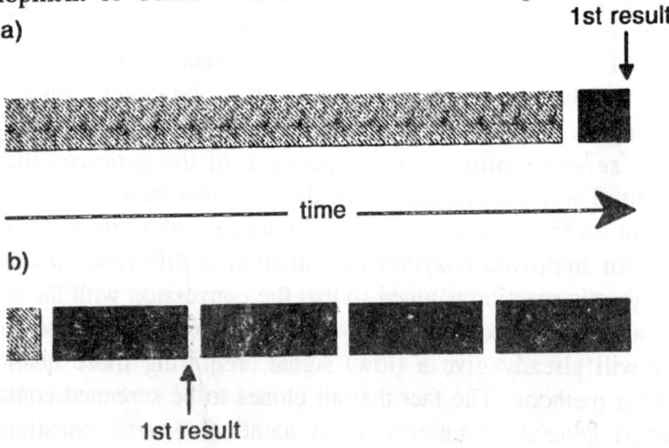


*Figure 3.2 : Graphical representation of the effect of a short integral screening time on the time 1st results are generated.*

Reader assays (both UV/Vis and fluorescence) are appropriate, in principle, for use in UHTS. A 96-well *microtiter* plate can be measured in less than 30 s, leading to a *theoretical* throughput of $10^4$–$10^5$ samples a day in the case of an endpoint measurement.

Unfortunately, this throughput can only be achieved with simple screening *protocols*, which are only possible when the conversion of substrate into product leads to strong changes in molar extinction coefficient or *fluorescence*. As this is quite often not the case with the actual substrates and/or products and use of *artificial* substrates is undesirable, derivatization of substrates and/or products into highly absorbing or *fluorescent* compounds is necessary as an additional step. These derivatization reactions not only lead to much more complicated screening protocols and consequently longer screening times (less throughput), but also to much longer development times as they require careful tuning of the *derivatisation* conditions (e.g. reagent concentration, temperature, pH, and reaction time) and solutions for *matrix* effects. Overall, this will result in a relative long integral screening time, which makes such a screening method less attractive for applications in CCM.

Figure elsewhere in this chapter shows a situation that is better suited for use in CCM. In this case, implementation of a screening method with very short development time and *sufficient* throughput (not necessarily ultra high throughput) generates the first screening results in a shorter period of time, leading to quicker hit identification and start of application research as well as sample preparation for the customer. The characteristics needed (short development time and sufficient throughput) are offered by *generic assays*, as for example based on nuclear magnetic resonance (NMR) or MS. These types of assays are characterised by the fact that they generate distinct signals for many chemical *functionalities* enabling straightforward differentiation between substrate and product of the *enzymatic* reaction to be screened. Combined with their relatively good *tolerance* to *fluctuations* in the matrix, these instrumental *assay* techniques are quite powerful.

A different approach is the proactive development of less *generic assays* for frequently requested chemical group conversions, as for example the synthesis of amino acids or for reactions producing *ammonia* as a sideproduct. In this case the challenge is to find reagents that are specific for these *molecules* and generate a clearly distinguishable signal in the derivatization reaction. Further more, this derivatization reaction must not be interfered by other chemical groups in these molecules, which will be different in each customer's request.

In the next sections of this chapter examples will be discussed from each of these CCM compliant screening approaches.

## CCM COMPLIANT SCREENING METHODS

As stated in the previous section, reader assays are characterized by a very short measuring time per sample. However, often this does not translate into a short integral screening time as the product formed in the screening reaction lacks a chromophore or fluorophore and consequently needs to be chemically derivatized before it can be detected in a reader with sufficient sensitivity and selectivity. In other cases the product of the screening reaction can be detected in a reader by a coupled enzymatic assay. In such an assay an enzymatic reaction cascade leads to the formation of a reader *detectable* compound (for instance $NAD^+$/NADH). In both approaches, the overall screening *protocols* are quite *laborious* and their development is often time consuming.

Nevertheless, reader assays can be successfully applied within the CCM context, for instance when the formation of the (side)products results in a strong change in molar extinction coefficient, or when the (side) products formed belong to a *category* of compounds for which a derivatization method or a coupled enzymatic assay has been developed beforehand.

### Optical Spectroscopic Methods Based on the Spectral Properties

The possibility of implementing screening methods based solely on changes in spectral properties upon conversion of substrate into product is small. Nevertheless, the great potential of this method justifies the efforts that are needed to investigate whether it is possible. Screening requests for transformations of molecules that contain a suitable chromophoric or *fluorophoric group* can best be started by recording the optical (UV/Vis/*fluorescence*) spectra of substrate and product. In this way it can be established whether the product absorbs and/or emits light of a specific *wavelength*. By including the appropriate blanks in these first experiments, interference of other compounds in the matrix (e.g. the reaction *buffer* and *cell* derived compounds) can be assessed. If these experiments have a positive outcome, a simple screening procedure can be developed in a few days.

*Example: Isolation of the D-p-Hydroxyphenylglycine Aminotransferase Gene*

One of the process concepts for the production of *enantiomerically* pure α-H-α-amino acids is the conversion of aketo acids to the desired amino acids by *aminotransferases*. For the production of the *denantiomers* of α-*Haamino acids*, the "*standard*" aminotransferases require equimolar amounts of D-Ala, D-Glu, or D-Asp as amino donor, which makes this process less attractive as these compounds are rather expensive. A more attractive *aminotransferase* process to *enantiomerically* pure D-α-H-α-amino acids is possible with an enzyme that can use cheaply available L-Glu and/or L-Asp as amino donor instead of their *D-congeners*. Such a stereoinverting aminotransferase had been described before in relation to the production of *D-p-hydroxyphenylglycine* and *D-phenylglycine*. These amino acids are produced as side chains in the manufacture of semisynthetic β-lactam antibiotics such as *cephalexin*, *cephadroxyl*, *ampicillin*, and *amoxycillin*. The reaction catalyzed by this *aminotransferase* (HpgAT).

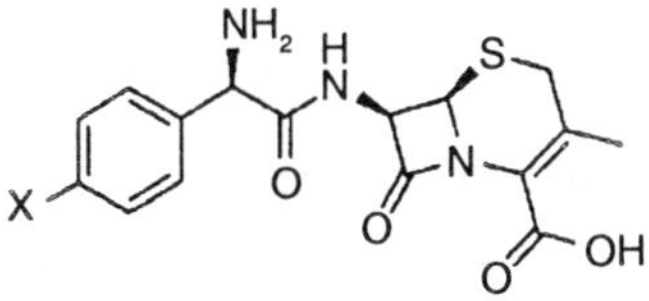

6: X = H
10: X = OH

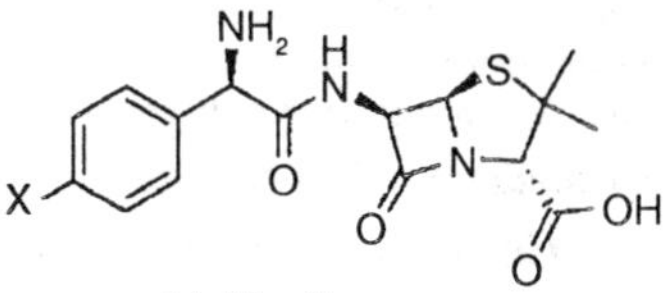

11: X = H
12: X = OH

We decided to identify the HpgAT gene by screening bacterial expression libraries including that of *Pseudomonas putida* NCIMB 12565, which had been reported to contain HpgAT activity. In this case neither a PCRbased approach nor the synthesis of the complete gene were possible, as the sequence of the HpgAT gene had not been published. Recording of the optical spectra of *D-p-hydroxyphenylglycine* and *p-hydroxyphenylglyoxylate* showed that the latter compound has a specific absorbance at about 300–350 nm.

Based on this observation, we designed the following simple screening strategy.

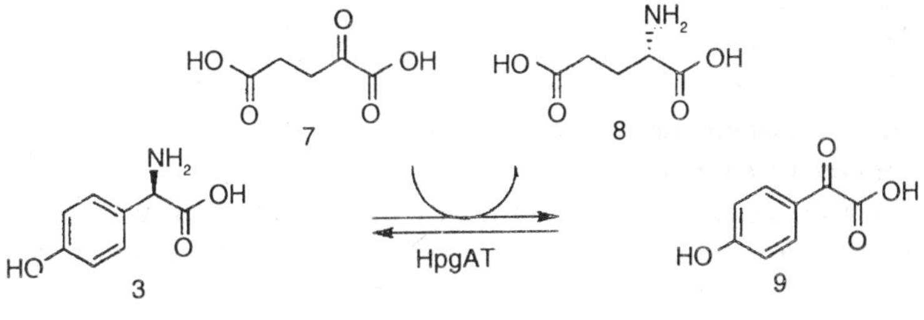

*Figure 3.3 : Reaction catalyzed by the stereoinverting D-p-hydroxyphenylglycine aminotransferase (HpgAT).*

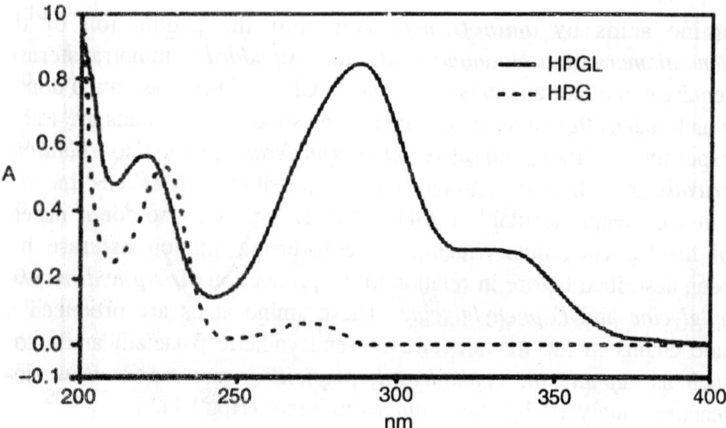


*Figure 3.4 : Optical spectra of* phydroxyphenylglycine *(HPG, dashed line) and* p-*hydroxyphenylglyoxylate (HPGL, black line) in 45 mM potassium phosphate buffer, pH 7.0. Concentration of both compounds was 55 μM.*

### *Procedure : UV Screening of Aminotransferase Reactions*

Colonies from a gene library were cultivated in microtiter plates containing 150 μL of a rich medium containing the appropriate *antibiotic*. After 16–20 h at 28°C, cells were harvested by centrifugation, washed once with 50 mM *potassium phosphate* buffer, pH 7.0, and resuspended in 180 μL reaction mix (100 mM *potassium phosphate*, pH 7.0, 15 mM *α-ketoglutarate*, 0.1 mM *pyridoxalphosphate* (PLP) and 0.5% v/v *Triton* X100. The screening reaction was started by addition of *D-p-hydroxyphenylglycine* to a final concentration of 5 mM, after which the $A_{340nm}$ in each well was monitored for 20 min using an Optimax microtiter plate reader.

As expected, *clones* containing the desired HpgAT gene showed significant increase in $A_{340nm}$ relative to negative wells, as can be seen in Figure elsewhere in this chapter Integral screening time in this specific case was less than one month due to the very short development time and modest analysis time (20 minutes per 96-well microtiter plate).

### *Optical Spectroscopic Methods Based on Followup Conversion of Product*

As the majority of the industrially relevant conversions are not accompanied by clear changes in optical signal, direct readerbased assays are frequently not an option. In many cases, however, a reader assay can still be developed by linking the actual screening reaction to a subsequent conversion that leads to a signal generating compound.

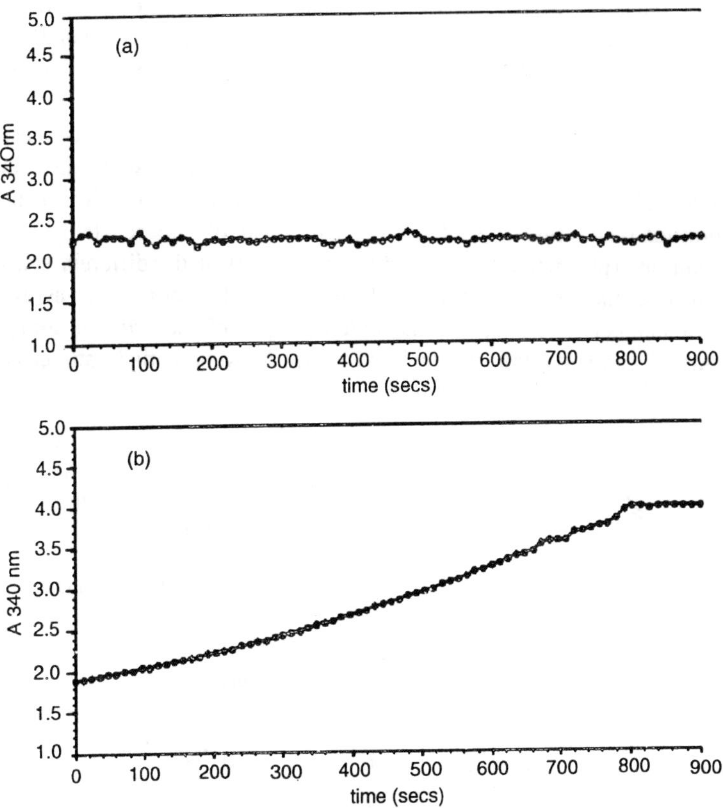


*Figure 3.5 : Typical progress curves of (a) a negative E. coli clone and (b) an E. coli clone containing the HpgAT gene.*

This can be realized in various ways. In a first approach, a chemical reagent or complexing agent reacts with a functional group of the (side)products. As these derivatisation reactions often require harsher conditions (especially pH) than the enzyme reaction to be screened, these types of as says are typically endpoint measurements. Interference of the measurement by other compounds that contain the same functional group as the product is the major complicating factor in these *derivatisation* based reader assays. A second approach is based on the subsequent conversion of one of the (side)products of the screening reaction in an enzymatic reaction cascade, in which the final step is a conversion of a readerdetectable compound (for instance $NAD^+$/NADH). The major advantage of this second approach is the higher selectivity of the enzyme compared with the chemical derivatization reagent, thereby strongly reducing the number of potentially interfering compounds. Furthermore, the mild conditions needed for the enzymatic cascade more often enable measurements in a kinetic mode.

Although possible in principle, the long development time of these kinds of optical *spectroscopic* assays hampers their application in CCM. This is due to the large number of parameters that have to be optimized and tuned, such as concentration of the reactants and conditions (pH, temperature, and reaction time) of the different steps as well as their order. *Proactive* development of robust reader assays for a number of frequently encountered types of *molecules* ("*group assays*") can solve this problem and makes these potentially attractive assays CCM timelines compliant.

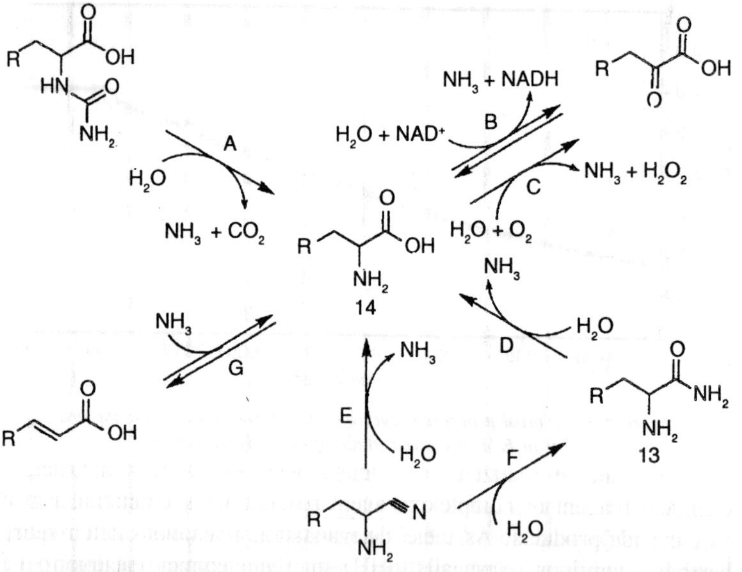

*Figure 3.6 : Schematic overview of different enzymatic reactions to enantiomerically pure amino acids. (A) carbamoylase; (B) amino acid dehydrogenase; (C) amino acid oxidase; (D) amidase, (E) nitrilase; (F) nitrilehydratase; (G) ammonia lyase.*

This approach is further elaborated in the following examples relating to synthetic routes for enantiomerically pure amino acids, which are important chemical building blocks for the pharmaceutical and agrochemical industry. A significant number of the CCM market requests fall within this *category*. A large number of different chemoenzymatic process concepts for the production of enantiomerically pure amino acids have been developed and commercialized, and *ammonia* is involved in the majority of these concepts either as sideproduct (*amidases*, *carbamoylases*, *nitrilases*, or *amino* acid *oxidases*) or as substrate (for instance when using amino acid *dehydrogenases* and *ammonia lyases*). We set out to develop reader assays for amino acids as well as for ammonia and primarily focused on amidases as these are excellent tools for the production of enantiopure α-H and α, α-disubstituted aamino acids.

During development of the assays for ammonia, we noticed consistently high backgrounds when crude extracts or cell suspensions of microorganisms were used. This is caused by the growth media, which often contain yeast extract and peptone. Growth of microorganisms in these media involves the production of ammonia from the peptides and amino acids, as these media contain a surplus of nitrogen over carbon. Therefore, it may be necessary to harvest the cells by centrifugation followed by washing with a suitable buffer before starting the actual screening in order to get a sufficiently low background.

*Example: Fluorometric Detection of Amidase Activity by o-Phthaldehyde/Sulfite Derivatization of Ammonia*

One of the reader assays we developed for the quantitative determination of ammonia is based on the derivatization reaction of ammonia with *o-phthalaldehyde* and sulfite, which results in a fluorescent derivatization product. The exact structure of this fluorescent product is not known.

$$NH_3 + \underset{15}{\text{(o-phthalaldehyde)}} + SO_3^{2-} \longrightarrow \text{fluorescent complex } (\lambda_{exc}\ 390\ nm / \lambda_{em}\ 538\ nm)$$

*Figure 3.7 : o-Phthalaldehyde/sulfite derivatization reaction of ammonia leading to a fluorescent product.*

We modified the original flowinjection method into an analytical procedure that can be run in an automated pipetting robot in combination with a microtiter plate fluorescence reader instrument. Using this hardware combination large numbers of enzymatic samples can be screened. The method shows considerable selectivity for ammonia over amino acids and amino acid amides, which makes the method also suitable for *monitoring* conversion in amidase reactions.

The *o*-phthalaldehyde/sulfite assay was successfully applied in the screening of an epPCR mutant library of a *bacterial* amidase gene. This screening targeted mutants with improved activity towards a specific aHaamino acid amide, without affecting this *biocat's* absolute *enantioselectivity* and good process stability.

*Procedure : Amidase Assay by Spectroscopic Ammonia Detection*

*E. coli* colonies containing the mutant library were cultivated in 96wells *microtiter plates* containing 150 μL of a rich medium supplemented with the appropriate *antibiotic* and *inducer*. After 18 h at 37 8C, 50 μL cell suspension was mixed with 50 μL of a 20% racemic a*Haamino acid amide* solution (pH 8). After 2 h of incubation at 55 8C, the reaction was stopped by adding 200 μL 1 M $H_3PO_4$, after which the microtiter plates were stored at 4 °C before *ammonia* was assayed. The stopped assay mixture and calibration solutions (*vide infra*) were diluted by mixing 20 μL with 250 μL 0.1 M $H_3BO_3$, pH 9.5. From this mixture, 10 μL was mixed with 45 μL of a 50 mM *o-phthalaldehyde* solution (in 25% methanol, 75% 0.53 M $H_3BO_3$, pH 9.5) and 45 μL 50 mM $Na_2SO_3$in a 384well *microtiter plate*. After 30 min at 55 8C, *fluorescence* was determined (excitation at 390 nm, emission at 538 nm) in a Fluorstar fluorimeter. As the signal obtained after this *incubation* at 558C still slightly increased in time, it was essential to measure all plates after the same derivatisation time.

For quantification it is a prerequisite that the calibration samples resemble the gene library samples as much as possible. We achieve this by imitating a *realistic amidase* reaction in the same biological *matrix* as present in the screened library samples (that is, as conversion proceeds we get increasing amino acid and ammonia concentrations and correspondingly *equimolar* decreasing amino acid *amide* concentrations). We call this a complementary calibration procedure. All calibration samples were prepared in 0.67 M $H_3PO_4$ and subjected to the same procedure as the mutant library samples.

To establish the standard deviation of this *screening* method, the whole *screening protocol* was executed with a few *microtiter plates* with *E. coli* cells containing the wildtype amidase gene. The standard deviation of the enzyme activity was less than 15%.

In this specific screening, 30 microtiter plates each containing 96 different random clones resulted in a dataset from 2880 *enzyme assays*. The absolute data values were corrected for the amount of cells in the assay by dividing the fluorescence by the optical density (OD600nm) of the cultures.

The resulting specific activities were compared with the activity of the wild type clones, which resulted in the identification of 41 possible hits for further testing. These mutants were rescreened on a somewhat larger scale. In this second round eight positive mutants were confir-med. Chiral high performance liquid *chromatography* (HPLC) analysis of the mutants proved that the *enantioselectivity* of the *amidase* was not lost in the mutants.

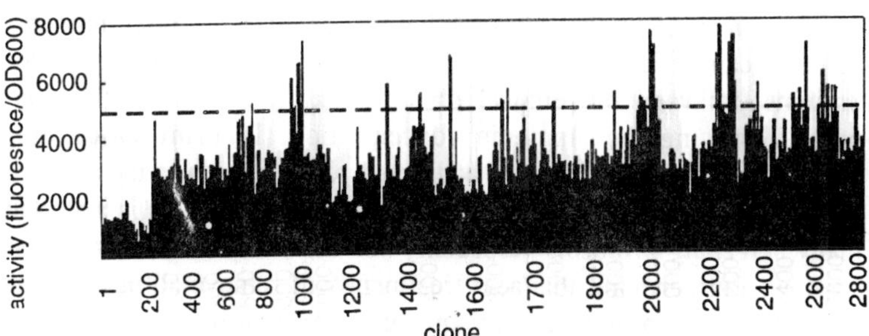


*Figure 3.8 : Data of an ophthalaldehyde/sulfite screening: activities of 2880 clones from an amidase epPCR library normalized for the cell density of the cultures ($OD_{600nm}$). Wildtype level is shown as a dashed line.*

### *Example: Colorimetric Detection of Amidase Activity by Detection of Ammonia via Glutamate Dehydrogenasecoupled Assay*

A second reader assay we developed to measure the ammonia concentration in highthroughput mode is based on the *enzymatic oxidation* of NADH by *glutamate dehydrogenase.* In this reaction *glutamate dehydrogenase* converts its substrates, *aketoglutarate* and *ammonia*, into L-Glu thereby oxidizing NAD(P)H to $NAD(P)^+$. The oxidation of NAD(P)H can be followed in a spectrophotometer at 340 nm or in a fluorimeter (excitation at 339 nm, emission at 460 nm). If the concentration of ammonia to be determined is lower than the concentration of *aketoglutarate* and NAD(P)H, the decrease in NAD(P)H is directly related to the concentration of *ammonia* in the sample. It can be calculated using the extinction coefficient of NAD(P)H at 340 nm, $6.22 \times 10^3 M^{-1} cm^{-1}$.

As *glutamate dehydrogenase* is very specific for ammonia, high concentrations of *amino acids* and amino acid amides do not interfere with the measurements. This makes this assay very suitable for the quantification of ammonia in samples from different enzymatic conversions as depicted in Figure elsewhere in this chapter, as long as neither $NAD(P)^+$ nor the reduced form are cofactors in such reactions. An other advantage of this method is the stable signal obtained after the glutamate dehydrogenase reaction has gone to completion. This makes the glutamate dehydrogenase incubation step less vulnerable than the *o-phthaldehyde sulfite* derivatization. Since screening projects for biocatalysts are usually carried out with *crude* extracts or cell suspensions of *microorganisms*, these suspensions may contain high NADH oxidase activity. This will lead to an overestimation of the ammonia concentration because NADH is also oxidized by these enzymes. To prevent this, NADH oxidase activity (as well as other disturbing enzyme activities that may be present in this ammonia as say), can be inactivated by acid treatment of cells, debris and/or proteins, followed by removal of the precipitate via centrifugation, before assaying the ammonia concentration. In many cases, this inactivation will not require an additional *pipetting* step as the addition of the acid also stops the actual screening reaction, which is a step in many *screening protocols* (especially for improved enzyme variants) anyhow. Furthermore, this acid treatment will remove all particulates from the screening reaction mixture, preventing noise in the *spectrophotometric assay*.

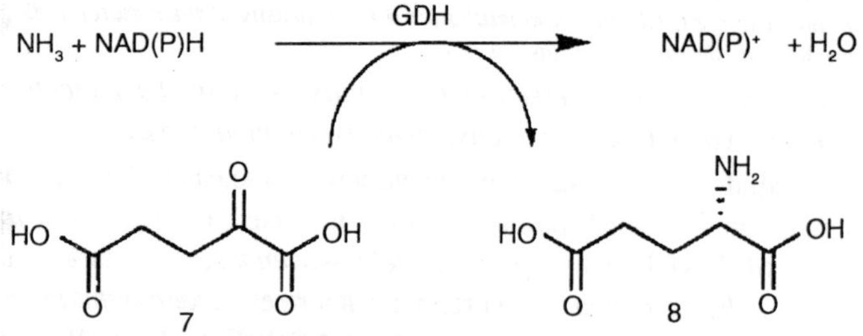

$$NH_3 + NAD(P)H \xrightarrow{GDH} NAD(P)^+ + H_2O$$



*Figure 3.9 : The glutamate dehydrogenase (GDH) reaction, coupling the amount of ammonia in a sample to oxidation of NAD(P)H.*

We used the glutamate dehydrogenase ammonia assay in a screening project to identify amino acid amide racemases (ARs). In conjunction with an *enantioselective amidase*, these enzymes enable the synthesis of enantiomerically pure α-H-α-amino acids from the corresponding racemic α-H-α-amino acid amides in a *dynamic kinetic* resolution (DKR) with 100% theoretical yield. Without ARs this amidase process is a *kinetic* resolution with an inherent maximum yield of the desired amino acid *stereoisomer* of only 50%.

The method for this screening for ARs is shown in Figure elsewhere in this chapter. Expression libraries in *E. coli* are incubated with a daHaamino acid amide as substrate, which is converted into the corresponding L-α-H-α-amino acid amide by ARpositive *E. coli* clones only. As the assay mixture also contains the strictly selective *laminopeptidase* from *Pseudomonas putida*, the L-α-H-α-amino acid amide formed is hydrolyzed to the L-α-H-α-amino acid under formation of ammonia, which will subsequently be determined using the glutamate

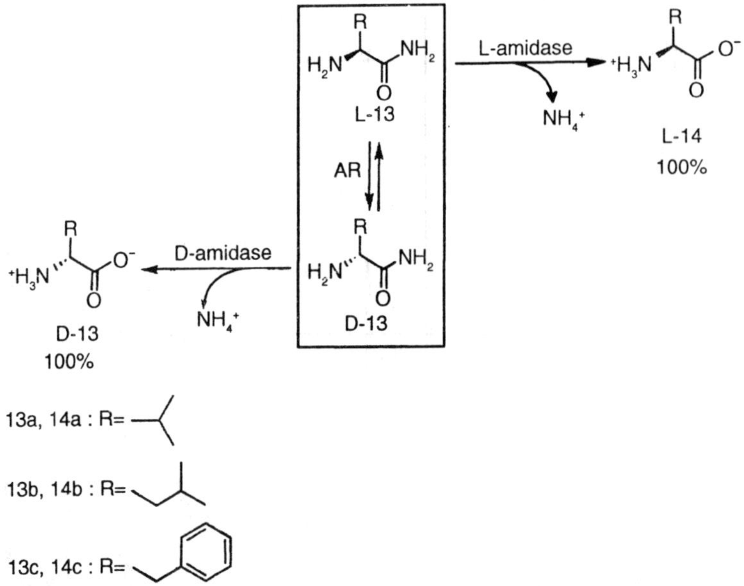

*Figure 3.10 : Dynamic kinetic resolution (DKR) of α-H-α-amino acid amides by combining an enantioselective amidase with an amino acid amide racemase (AR), leading to 100% theoretical yield of the desired amino acid enantiomer. The AR reaction is boxed.*

*dehydrogenase assay*. As can be seen in Figure elsewhere in this chapter, this screening approach will not only identify ARpositive clones, but also dselective or nonselective amidasecontaining clones, since they will directly hydrolyze the D-α-H-α-amino acid amide screening substrate. Therefore, more specific followup assays were needed to distinguish damidase from ARpositive clones. This was done on an HPLC system in which all four components of the reaction mixture, namely the α-H-α-amino acid amide and α-H-α-amino acid enantiomers, were quantified after derivatization with *o*phthaldehyde in combination with *D-3-mercapto-2-methylpropionic* acid as chiral reagent.

*Procedure : Enzymecoupled Assay for Amino Acid Amide Racemase via NADH Detection*

The actual AR screening was executed as follows. Bacterial expression libraries in *E. coli* were cultivated in 96-well microtiter plates containing 200 μL of a rich medium with the appropriate antibiotic. The cultures were grown at 700 rpm and 28–30 8C for 2 days, before cells were harvested by centrifugation (10 min at 1500

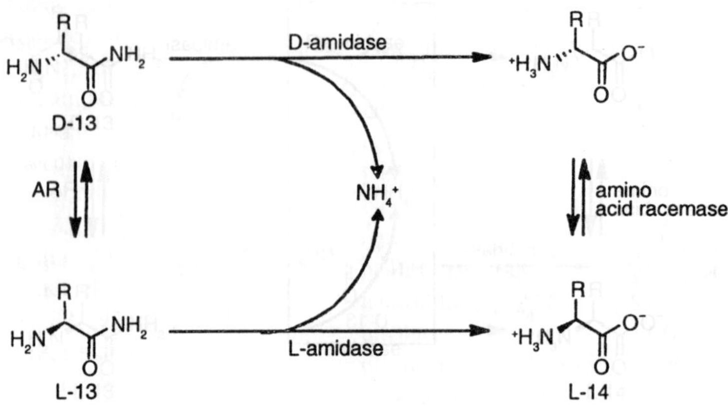


*Figure 3.11 : Schematic representation of possible reaction pathways using a D-α-H-α-amino acid amide as screening substrate for an amino acid amide racemase (AR).*

× *g*), washed twice with 50 mM Tris–HCl buffer, pH 7.5, and resuspended in 50 *u*L of 50 mM Tris–HCl buffer, pH 7.5. The screening reaction was started by the addition of 50 μL of substrate solution containing a mixture of 140 mM *d-valine amide*, *d-leucine amide* and *d-phenylalanine amide* each in 50 mM Tris–HCl buffer, pH 7.5 and 2 mM $MnCl_2$. After 20 h incubation in a *microtiter plate* shaker at 30°C and 700 rpm, 2 μL of help solution containing the *laminopeptidas*e was added to be sure that all L-α-H-α-amino acid amide formed in the screening reaction and not yet converted by the endogenous *E. coli aminopeptidase*, was converted into *ammonia* and L-α-H-α-amino acid. After an additional 1.5 h incubation at 30°C, all reactions were stopped by the addition of 100 μL of 0.15 M HCl. After removal of the cell debris by centrifugation (10 min at 1500 × *g*), 7 μL of the clear supernatant was transferred to new *microtiter plates* containing 93 μL of *glutamate dehydrogenase* (GDH) reagent per well. This GDH reagent contained (per 100 mL) 43 mg of NADH, 116 mg of *α-ketoglutarate*, 11.8 mg of ADP and 1200 U of GDH in 150 mM Tris–HCl buffer, pH 8.0. After 15 min incubation at 37°C, potential positive clones were determined by measuring the $A_{340nm}$ in each well using a *Spectramax* 250 plus microtiter plate reader. The $A_{340nm}$ values obtained in the AR screening of the *Ochrobactrum anthropi* IA gene library are shown in Figure elsewhere in this chapter.

In order to reduce *platetoplate* variation within the $A_{340nm}$ data, the following calculation method was used to identify positive hits. From each 96 data points from a 96-well *microtiter plate* the average A340nm and standard deviation were calculated.

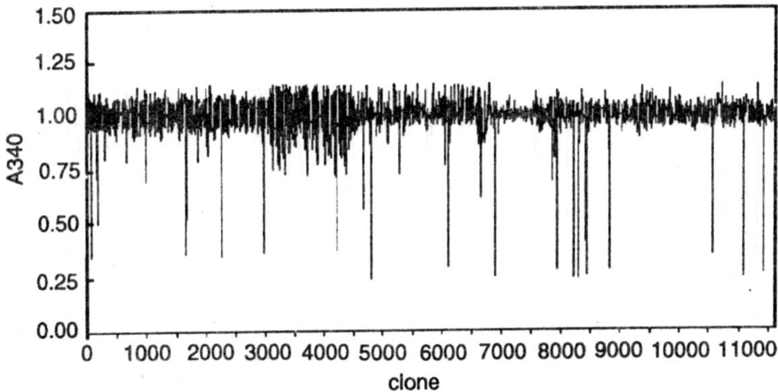


*Figure 3.12 : Data of a glutamate dehydrogenase screening: decrease of A340nm of 11000 clones from a gene library of O. anthropi expressed in E. coli. Absolute vaiues obtained have been normalized to the median of each microtiter plate.*

Subsequently, the standard deviation was multiplied by a factor of 3 and subtracted from the average. The resulting number was considered as the *cutoff value*. Clones with an $\Lambda_{340nm}$ lower than this cutoff value were regarded as potentially positives and selected for further screening. In this way the effect of random noise and platetoplate variation was diminished.

Based on this criterion, 32 potential positive clones were selected out of the 11 272 clones screened. In the followup assays, 22 of the 32 potential positives showed clear α-H-α-amino acid formation, one of which contained the desired AR gene, as this clone clearly showed lleucine formation from *d-leucine amide*. The other 21 clones expressed no AR but were most likely either dselective or nonselective amidases.

### *Example: Colorimetric Detection of Amino Amidase Activity Using $Cu^{2+}$as Sensor for Amino Acids*

Besides assays based on ammonia detection, we also developed reader assays for the detection of reactions involving amino acids. One of the requirements for such an assay was that it had to be applicable in screening of *amino amidases* (amidases requiring an aamino group for activity). This implied that formation of a small amount of α-amino acid (product) in a matrix with a usually much higher concentration of aamino acid amide (substrate) had to result in a clearly detectable change of spectral or *fluorescent* properties, in spite of the fact that both compounds contain a primary amine group. Furthermore, interference of the assay by ammonia, which is formed as sideproduct in equimolar amounts, had to be excluded.

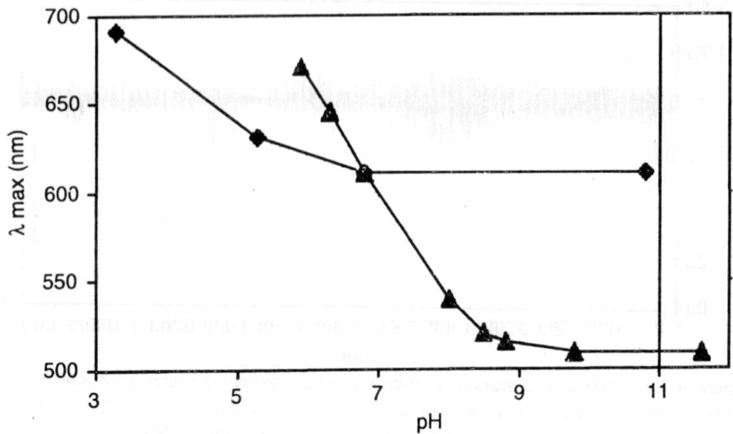


*Figure 3.13 : Influence of pH on $\lambda_{max}$ of the $Cu^{2+}$complex of valine amide (triangles) and valine (diamonds).*

Recently we described a method for the determination of amino amidase activity that meets these requirements. This method is based on the use of $Cu^{2+}$alone as metal sensor, which distinguishes this method from earlier methods that use $Cu^{2+}$or $Ni^{2+}$as metal sensors in conjunction with a fluorescent *macrocyclic metal chelator*. It appears that the copper complexes of an *α-amino acid* and an *aamino acid* amide have different spectral properties at al kaline pH.

Below pH 8, α-amino acid amides form bidentate complexes with $Cu^{2+}$, wherein two CuO and two CuN bonds are present, resulting in a blue complex ($k_{max}$ 620–680 nm). At a pH higher than 8, a violetred complex is formed ($k_{max}$ 510 nm). The structures of both complexes (the blue as well as the *violetred* one) have been described before. The $Cu^{2+}$complex of *α-amino acids*, on the other hand, stays blue upon a pH change from 3.3 to 10.8, although there is a shift in $k_{max}$ from 690 to 620 nm. The above observations were the basis for the newly developed amino *amidase assay* method, as this activity will then be indicated by a color change from violetred (aamino acid amide) to blue (α-amino acid) provided that the pH of the color assay mixture is higher than 8.

We successfully applied this assay method in monitoring amino amidase activity in a large number of fractions resulting from enzyme purification studies. *α-Methylvaline amide* was used as substrate, which was converted into a *methylvaline* by the desired *amino amidase*.

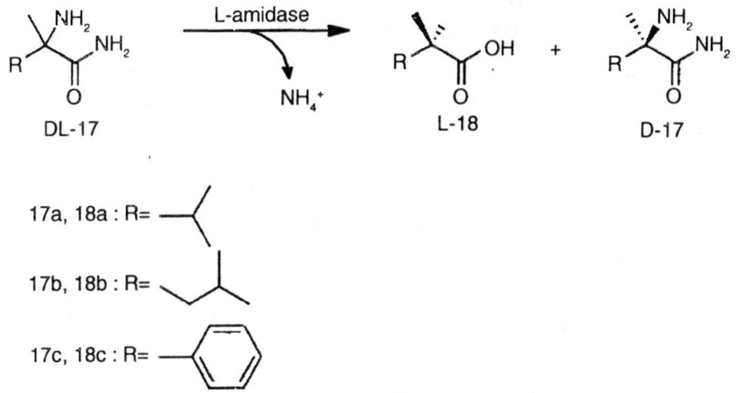

*Figure 3.14 : Conversion of DL-α, α-disubstituted α-amino acid amides by an Lamidase.*

*Procedure : Colorimetric Amino Amidase Assay Using $Cu^{2+}$*

An aliquot of 50 μL of each fraction was mixed with 100 μL of a 130 mM dlamethylvaline amide solution, pH 8.0, and 50 μL 400 mM HEPES.KOH, pH 8.0 buffer. After 1 h incubation at 37°C, the *amino amidase* reaction was stopped by the addition of 100 μL 1 M $H_2SO_4$. After removal of particulates by centrifugation, 70 μL of the clear stopped reaction samples were added to 74 μL of an *aqueous* 20 mM $CuSO_4$ solution followed by the addition of 56 μL of a 43 : 5 mixture of 0.4 M borate buffer pH 9.4 and 10 M KOH. After mixing for 10 s, the absorbance was determined at 620 nm in an Optimax microtiter plate reader (Molecular Devices).

Quantification of the amethylvaline formed was possible using a complementary calibration curve approach. In this specific case

such curve was prepared by mixing increasing amounts of a solution containing 130 mM ammonia and 130 mM *methylvaline* with correspondingly decreasing amounts of a 130 mM *amethylvaline amide* solution (total concentration amino acid and amino acid amide: 130 mM). Just before measurement, 220 μL of each calibration solution was mixed with 220 μL 1 M $H_2SO_4$and 220 μL 0.2 M HEPESKOH, pH 9.0 to obtain the identical matrix as in the stopped reaction mixtures.

The results obtained showed that fractions containing the desired amino amidase were easily distinguishable by this simple *spectrophotometric assay*. Furthermore, validation of this $Cu^{2+}$method by measuring the amino *amidasecatalysed* conversion of *amethylvaline amide* by this method and HPLC/MS showed that the reader assay delivers quantitative data.

## CCM COMPLIANT SCREENING METHODS BASED ON GENERIC INSTRUMENTAL ASSAYS

The short integral screening time required in CCM can also be met by use of generic assays. These assays, which are mostly instrumental, combine extremely broad applicability with short development time. Therefore, the introduction of these methods can in most cases be done reactively after receiving a question from a customer. This clearly distinguishes them from *assay* methods requiring proactive development to be compliant with CCM timelines, as is the case with reader assays in which chemical derivatization or *enzymatic cascade* reactions are needed to visualize the product. *Generic* instrumental assays, for instance, are based on NMR and MS.

### Flow-injection NMR as Analytical Tool in Highthroughput Screening for Enzymatic Activity

NMR spectroscopy is a wellestablished analytical technique that can yield structural and quantitative information about molecules present in solution. Originally, it was mainly applied in the synthetic field and for *physicochemical research*; nowadays there are numerous new applications in biology and medicine. The historical drawbacks of low sensitivity and overlapping signals have been overcome by the use of higher field strengths with their intrinsically higher sensitivity and better dispersion of resonances. Recently, the advent of so called *cryoprobeheads* again increased the sensitivity by a factor of four.

Minimal sample pretreatment and better capability of water suppression also make samples of *biological origin amenable* for routine analysis. In *enzymatic* conversions substrate and product are usually different chemical entities with different *magnetic* properties, so they yield different NMR spectra. Due to the quantitative nature of NMR *spectroscopy*, conversions can easily be *monitored*, measuring both *substrate* and product.

The availability of classical sample changers makes recording of about 100 spectra per day possible. To increase the number of measurements per day, different methods of sample introduction have been developed and the application of flowthrough probes has increased the number by one order of magnitude, so that$^1$HNMR *spectroscopy* can now be applied to medium/highthroughput screening. In the literature, screening in drug discovery and development, com *binatorial chemistry* libraries and *body fluids* is well described.

#### *History*

Convinced by the potential of NMR *spectroscopy* in the analysis of large series of similar samples, until then limited in our practice to 60 per night, we were thinking about increasing the throughput by alternative sample introduction techniques. *Liquid Chromatography/* NMR (LC/NMR) had been developed in the late 1980s and further developed in the 1990s and the use of sample loops made it possible to store HPLC fractions in those loops and later introduce and measure stored fractions in a flowthrough probe. In 1995 we carried out a first trial at the *Bruker site* with biological samples, manually introduced into the sample loops. Using the software developed for LC/NMR, it turned out to be possible to measure those samples sequentially. Although the speed at that time was still limited due to practical problems of the equipment (thin *capillaries* restricting the flow), the test was successful in that it proved to be possible to *automatically* introduce and measure NMR spectra of samples for which the only *treatment* had been the addition of some $D_2O$ as a lock solvent.

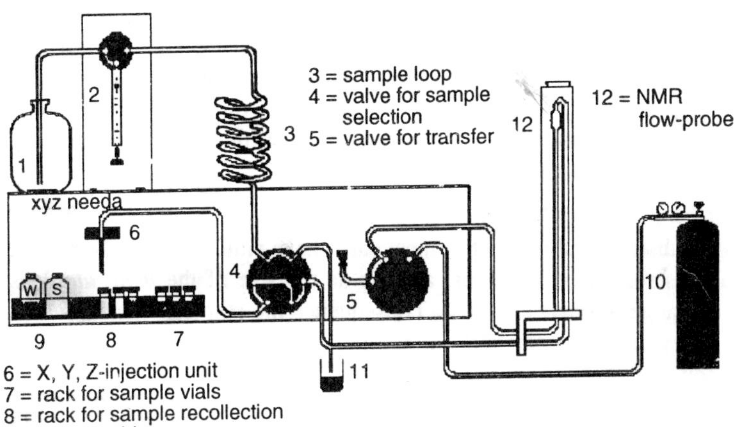


*Figure 3.15 : Schematic overview of the BEST NMR system.*

Bruker biospin continued the development, supported by our reallife samples and suggestions. They published together with the Nicholson group in open literature about the principle in 1997 and soon thereafter commercialised their product (Bruker Efficient Sample Transport (BEST) NMR) using a Gilson 215 as a liquid handler and sample changer for vials and well plates.

Our first flowthrough system (500 MHz) in 1998 was able to change and measure a sample in 3.5 min, which meant a total of 400 samples per day, already 4 times more than the classical sample changers could deliver. Further improvement, by optimising each step in the process, turned out to be possible in the years thereafter to about 1 min per sample. A crucial element in this was the use of a heatable transfer capillary, which, after careful calibration, made it possible to get the sample to the desired temperature before introduction to the flowcell, so that the normal temperature equilibration of 4–5 min could be skipped and the measurement started immediately.

For a more detailed description of the principles of flowthrough systems on the market the reader is referred to the review of Stockman et al.. A *schematic* overview of the BEST NMR setup based on the Gilson XL215 *autosampler* is given in Figure elsewhere in this chapter.

***Current Practice***

There are several reasons why NMR, and in particular flow-injection NMR, is suitable for screening large libraries for new or improved biocatalysts. These are explained below.

*Direct Measurement of Both Substrate and Product*

No artificial substrates are needed for detection purposes. As indicated earlier, this prevents erroneous conclusions, as the artificial substrate is not representative of the aimed substrate, especially in studies to improve the catalytic properties of a *specific enzyme*. Derivatization reactions are also unnecessary, thus obviating the need for additional steps to the screening.

*Intrinsically Quantitative*

Substrate and product in most cases have similar NMR properties, such as relaxation behavior, as long as the size of the *molecule* does not change dramatically. As a consequence the response factor, which is unity under controlled conditions, also applies under screening conditions.

*Short Development Time*

Finding out whether a particular (*enzymatic*) conversion can be detected by NMR is easily done by measuring the spectra of the intended *substrate* and *product* under the expected conversion conditions. This will take less than 1 h. If successful, the actual screening can start, if not, conditions like pH or *temperature* can be changed to see if the desired separation of the resonances can be obtained. In general it is not necessary to separate all the resonances of the substrate and the product, since one is sufficient.

Where no product is available yet, simulation of the NMR spectrum by specific computer programs gives an indication whether the screening in principle would be successful. In those *screenings*, manual inspection of the individual spectra is needed to find actual positive hits. This is easily done by *scrolling* through the spectra at the expected region of interest.

*No Calibration Necessary*

Because substrate and product generally have similar molecular weights, integrals of substrate and product resonances can be compared more or less quantitatively, without applying the normally needed long relaxation delay (about 30 s between scans). This way, the actual measuring time can be limited and in practice can be made the same as the time needed for the preparation of the next sample.

*No Critical Sample Pretreatment*

The standard sample pretreatment is centrifugation, sampling the supernatant with a robotized system (e.g. 350 $\mu$L) into a deepwell plate, and adding a constant amount (50 $\mu$L) of a solution containing $D_2O$, EDTA (50 mM), and a known amount of maleic acid. In this way, a lock signal is supplied, and the metal ions, present from the cellfree extracts, are chelated to prevent broadening of the resonances. Moreover, addition of this solution supplies an internal standard to monitor unexpected disappearance of the substrate. In the case of screening for new biocatalysts, it is generally not necessary to stop the screening reaction, for any positive hit is welcome, irrespective of the response in a specific incubation time. When the screening has been set up to improve the enzyme, the reaction should be stopped, for example by adding acid or base, at a moment that the reaction of the wildtype enzyme is around 10% of conversion. In that case improvements can easily be detected.

*Relatively Matrixindependent*

NMR spectra are in general not very sensitive to small changes in sample conditions. Most substrate incubations are done between pH 5 and 7, where the influence of pH on the NMR spectrum is normally negligible. If the reaction is stopped by the addition of acid or base, care must be taken that the resulting pH is far enough from the $pK_a$ values of the ionizable groups in the molecule, that is below pH 1 or above pH 12.

*No Reference Material Necessary*

To observe the conversion of a substrate to an expected product, it is best to have a sample of the product available. If this is not possible, as pointed out under "short development time", the spectra have to be inspected visually to find the resonances of the product. Once a positive hit has been found, the resonance positions of the product are known and the *remainder* of the *spectra* can be evaluated automatically.

*Screening for Multiple Substrates/Products Possible*

With flowinjection NMR it is possible to screen more than one substrate to be converted at a time. The boundary condition for this is that each substrate and product has at least one signal that does not overlap with the others, so that conversions for each substrate can be calculated. By measuring the individual spectra from potential substrates, clever combinations can be chosen. In our practice, up to four different substrates have been analyzed for conversion in one round of screening, saving time and effort needed for both the substrate in cubation and the measurement. The template for the calculation of the conversions can easily be adapted to report the conversions for individual substrates.

*Unexpected Conversions Detectable*

Another advantage of NMR spectroscopy is that, being a heuristic technique *par excellence*, it is not only the expected conversion that can be measured, but also conversions that were not intended. These can be conversions of other functional groups present or followup reactions of primary products.

*Robust*

NMR spectroscopy in general is a technique without direct interaction between the sample and the instrument during the measurement. There is no wear and no direct contact between the sample and measuring device. In practice, therefore, the downtime is very low. As long as the cryogenic liquids are replenished, the instrument will be in operation. If an electronic device fails, replacement with that item from the manufacturer's stock will solve the problem. For the flow-injection NMR system a specific area that can be *problematic* is the flow through the *capillaries* and the *flowcell*. As the inlet capillary is thinnest (0.5 mm diameter), possible clogging occurs either there or in the injection port. In those cases, reversal of the flow and flushing with sufficient amounts of cleaning liquid will help. Taking care that centrifugation of the microtiter plates is done correctly and the *injection needle* is adjusted correctly (so that the needle does not touch the debris present) will in practice prevent *clogging*.

***Practical Aspects***

To operate the flowthrough system routinely, a number of issues need to be dealt with.

*Temperature Equilibration*

As most enzyme assays are waterbased, a number of spectroscopic problems associated with NMR measurements in water (i.e. temperature equilibration and suppression of the huge water resonance) have to be solved. The temperature of NMR probes is normally actively stabilized at a few degrees above room temperature. As a consequence, a new sample from the sample changer has to equilibrate from room temperature to the temperature of the probehead. This takes about 4–5 min. By *preheating* the sample on the fly before entering the flowcell, this time can be saved, thus increasing the number of samples measured. This preheating is done in the capillary between the liquid *handler* and the *probehead*. Bruker developed a heatable transfer capillary that, after careful calibration, does the job. This calibration must be done with actual samples and using the intended dispensing speed on the fly of a series of measurements. When the *calibration* is correct, the sample will be at the desired temperature and the measurement can start immediately after sample introduction.

*Adjustment of Field Homogeneity*

In contrast to classical NMR tubes, the flowcell has a constant volume and wall thickness. Furthermore, the *measuring coil* is fixed to the outside of the cell. Consequently, each sample experiences the same surroundings, which results in an equal *homogeneity* of the *magnetic* field for all samples. In practice this means that the magnetic homogeneity after initial adjustment does not need to be changed or at most by minor adjustment of the Z and Z2 shims. These minor adjustments can be done during the dummy scans (scans without data acquisition, needed to acquire a steady state).

*Water Suppression*

Water suppression in automation is not trivial. Success is dependent on shimming and stable temperature. The advantage of screening with flowinjection NMR is, as mentioned earlier, that samples are very similar and that the environment inside the probe is more or less constant. In practice this means that initial shimming and determining the correct offset is sufficient to reach a good level of water suppression. Software *manipulation* of the data can cosmetically improve the spectrum, so that *automated processing* is possible. If resonances of interest are (too) close to the water resonance, increasing or decreasing the sample temperature will shift the residual water resonance to a more favorable position.

*Sensitivity and Detection Limit*

The detection limit on a Bruker 500 MHz machine is in the higher micromolar range. By taking more scans this can be lowered, but then the throughput rapidly decreases. When screening for new biocatalysts, substrate concentrations are used that mimic industrial application conditions whenever possible, which corresponds to concentrations of more than 50 mM. That means that in the case of a conversion of 1%, 0.5 mM should be *detectable*, which in practice is possible. Only when the resonances of the product are highly overlapped by those of background compounds, conditions should be optimized (pH, *temperature*) to allow detection.

*Back Mixing*

An essential prerequisite for applying flowinjection NMR in general is that samples do not mix in the flowcell. Separation of samples is accomplished by adding cleaning segments and small gaps with air to the sample train that is injected. The amount of back mixing is influenced by factors such as aspiration and dispensing speed, amount of cleaning liquid, number and size of trails, and composition of sample. Some samples can be very sticky, such as solutions containing sugars and polyols. Trial and error with different conditions will allow optimization of the method to be used. When screening large libraries (e.g. 10 000 samples or more) for a specific enzymatic activity, speed is limiting. Two factors balance the overall speed possible: sample preparation and data acquisition. The total ac quisition time should be equal to the time needed for the preparation of the next sample. This time plus the time that is needed for the actual dispensing of the sample then determines the repetition time. In the case of sufficient sensitivity, sample preparation and sample dispensing become limiting. However, only a few positive hits at most are to be expected and mixing becomes less critical and some mixing can be tolerated. Even 10% of mixing is acceptable, because samples will then report 10% of the activity of the previous sample following a hit and can therefore easily be recognized. An *automatic routine* in a *spreadsheet* program can also correct for carryover from the previous experiment.

Samples can be recovered, but this limits the speed to an unacceptable low level for screening purposes.

*Automation (Data Acquisition)*

To handle a large number of samples, the NMR manufacturers provide standard software so that the operation of a flow probe does not differ from that of a classical sample changer.

*Automation (Data Processing and Handling)*

To keep track of the fast acquisition of NMR data, further processing of the data (Fourier transformation, phase correction, baseline correction and integration) needs to be completely automated. Modern NMR software supplies commands that can do automatic phase and baseline correction that can also be implemented in macros. One method is to run such a macro each time a measurement has finished, the other, more reliable way is to determine the correct phase settings for the first experiment of a series (for instance a microtiter plate) and to run the processing for the series with these settings after the measurement of a *microtiter plate* has been finished. *Automatic baseline* correction is easiest when, after initial baseline correction of the complete *spectrum*, selected regions of interest are corrected with optimized *parameters* for a specific type of region. Integrals can be set automatically once they have been determined for a single experiment. In practice the settings are constant for all similar experiments.

To calculate conversions, *integrals* are put into a text file, which can then be imported into a spreadsheet on a separate computer. There, calculations can be performed by *macros* that can easily be adapted for individual experiments. For evaluation purposes, a *graphical* representation of the conversion in a microtiter plate and a list sorted on highest conversion can be added to a report.

*Cleaning*

As crude enzyme preparations contain some residual *biomass* even after *centrifugation*, *precautions* have to be taken to prevent *microbial growth* in the flow system. Adding *sodium azide* to the cleaning and pushing *liquids* and *flushing* the last sample with one or more dummy samples with cleaning liquid is sufficient for routine operation. Regular cleaning with surfactant solution, *alcohol*, *ammonia*, and *diluted nitric acid* is done to remove traces of biomass and sticky components.

*Capacity*

NMR machines can run more or less unattended for weeks, as long as the regular filling intervals for liquid *nitrogen* and *helium* are maintained. The Gilson 215 liquid handler can contain 10 microtiter plates and most of the screenings are done in 96-well format. With a repetition time of 1 min, new samples have to be added every 16 h, which during the weekends has to be carried out four times when large screenings are done. Switching to a 384-well format, composed of four 96-well plates, limits this number to one at most. *Deepwell plates* of this format can contain 360 $\mu$L liquid, which is sufficient for operation with a 4 mm flowcell. The use of a 3 mm *cryo flowcell* can be done with 160 $\mu$L and can thus limit the amount of sample needed.

*Example: Screening of a Bacterial Expression Library for Amidasecontaining Clones*

As well as α-H-α-amino acids, a,adisubstituted aamino acids are also a group of industrial relevance, as is exemplified by the use of lamethyl3,4dihydroxyphenylalanine (*lamethyldopa*) as an antihypertensive drug and *amethylvaline* as an intermediate for the *herbicide Arsenal*, currently marketed as its *isopropylamine* salt by BASF, and related *herbicides*.

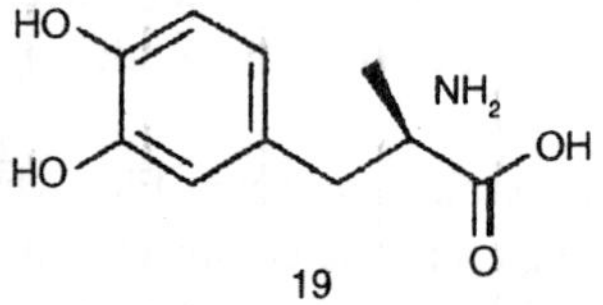

19 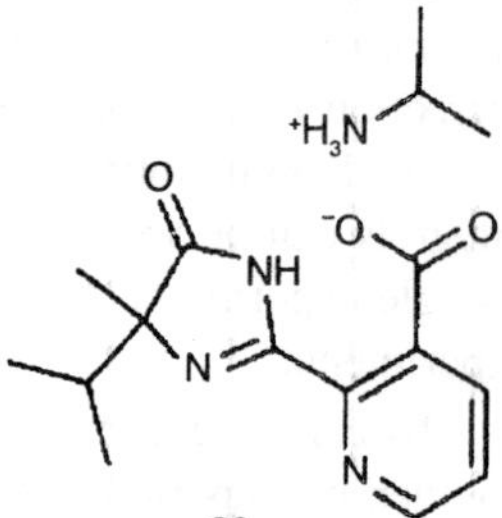 20

These disubstituted amino acids can be produced in *enantiomerically* pure form by an amidase process that is quite similar to the process for the production of α-H-α-amino acids. In this case, however, ketones are used as starting material instead of aldehydes and, since the *laminopeptidase* from *P. putida* ATCC is not active for this class of amides, a different amidase is applied in the resolution step.

The screening assay presented here was intended to identify new amidases with activity towards a,adisubstituted aamino acid amides. For this purpose, a methyl leucine amide was used as substrate, which will be converted into *amethyl leucine* by positive clones. The actual gene library screening was performed as follows.

*Procedure : Amidase Screening by NMR*

96-Well microtiter plates filled with rich medium (200 $\mu$L per well) supplemented with the appropriate *antibiotic*, were *inoculated* with transformants from a *bacterial gene* library in *E. coli*. After incubation for 48 h at 22°C at a shaking speed of 600 rpm, 100 μL of culture was transferred from each well to a wide well plate containing 300 μL dlamethyl leucine amide (67.5 mM, pH 8.0) and incubated for 24 h at 40°C. Subsequently, the *microtiter plates* were centrifuged for 10 min at 1000 × *g* to pellet the cells. From the supernatant in each well, 300 μL was transferred to deepwell 96-well microtiter plates each containing 50 μL $D_2O$ with 40 g $L^{-1}$ maleic acid, 20 g $L^{-1}$ EDTA, and 1 g $L^{-1}$ DSS (2,2-dimethyl-2-silapentane-5-sulfonate sodium salt) (pH of this solution was adjusted to 6.5). $^{1}$HNMR measurements were performed at 500 MHz on a Bruker Avance *spectrometer* using a 4 mm flowcell (active volume 120 lL) equipped with standard Bruker software (XWINNMR version 3.1 and ICONNMR 2.6). Aspiration and dispensing speed were set to 8 mL $min^{-1}$ and the sample volume was 300 μL.

The total measurement time was 25 s (eight scans with two dummy scans), equalling the simultaneous preparation time of the next sample. The overall repetition time was about 60 s. Processing (Fourier transformation, phase correction, baseline correction, and integration) was done per microtiter plate, using parameters optimized for the first sample of the *microtiter plate*. After a quick visual check of the spectra, the integral values were exported to a text file, which was imported into an Excel file, designed to calculate the conversions.

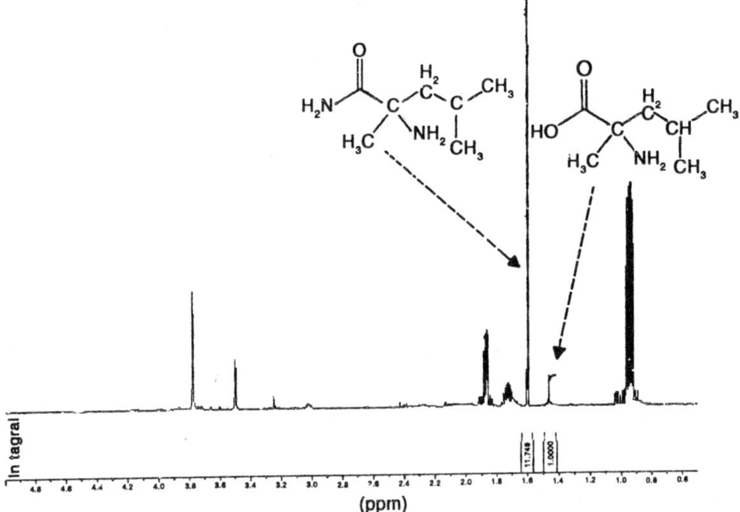


*Figure 3.16 : $^{1}$HNMR spectrum of a positive hit in the screening for an amidase that hydrolyzes DL-α-methyl leucine amide.*

For each microtiter plate, conversion figures and graphical representation of the results were displayed and plotted.

In total 120 microtiter plates were screened in this way, which led to the identification of two positive clones. A spectrum of one of these positive *clones* is given in Figure elsewhere in this chapter.

*Example: Identification of a Phenylpyruvate Decarboxylase Clone*

*Phenylpyruvate decarboxylases catalyze* the *decarboxylation* of *phenylpyruvate* resulting in the formation of *phenylacetaldehyde*.

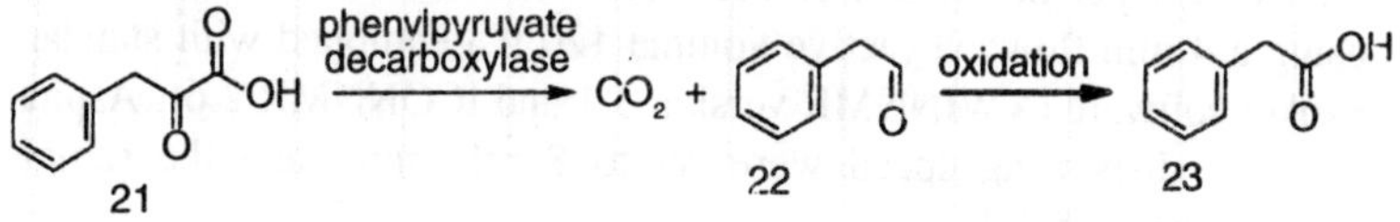

$$\mathbf{21} \xrightarrow{\text{phenylpyruvate decarboxylase}} CO_2 + \mathbf{22} \xrightarrow{\text{oxidation}} \mathbf{23}$$

*Figure 3.17 : Natural reaction catalyzed by the enzyme phenylpyruvate decarboxylase.*

Interestingly, this enzyme can also catalyze the *asymmetric acyloin* condensation of *phenylpyruvate* with various *aldehydes*. In a similar setup as that used in the amidase screening, but with 120 mM *phenylpyruvate* as substrate, a bacterial gene library in *E. coli* was

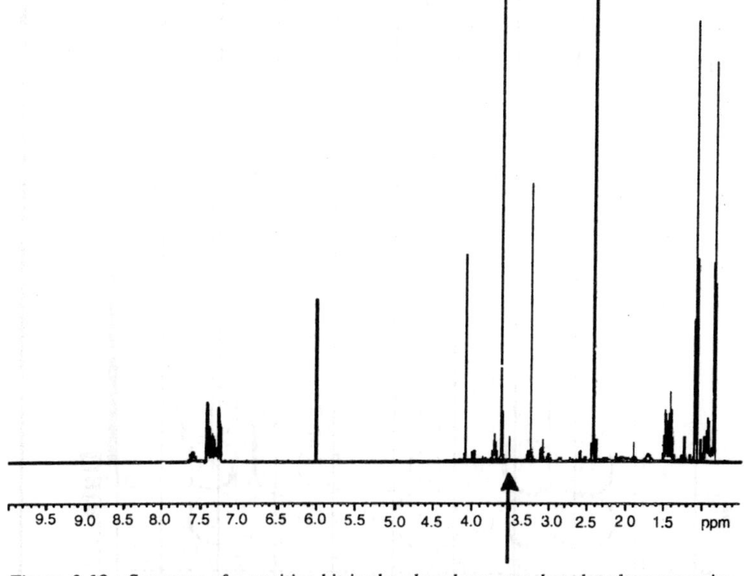


*Figure 3.18 : Spectrum of a positive hit in the phenylpyruvate decarboxylase screening. This spectrum indicates the formation of phenylacetic acid (resonance indicated by the arrow) rather than phenylacetaldehyde, most likely because of (chemical) oxidation of the screening product formed.*

screened and assayed with flow-injection NMR to isolate the phenylpyruvate *decarboxylase gene* from this microorganism. On visual inspection of the spectra it turned out that there were no hits, as formation of *phenylacetaldehyde* could be demonstrated in none of the wells. However, a few spectra could be found where the oxidation product of this compound (*phenylacetic acid*) was detected. Apparently, the primary *aldehyde product* had been oxidized into the carboxylic acid. Further examination of these clones by nucleotide sequence determination followed by subcloning proved that these hits did indeed contain the desired phenylpyruvate *decarboxylase gene*.

*Example: Identification of Amidase Mutants with Improved Activity towards aMethylphenylglycine Amide*

Using a combination of different directed evolution techniques such as epPCR and saturated mutation primer PCR DNA shuffling, the specific activity of the lamidase from *Ochrobactrum anthropi*

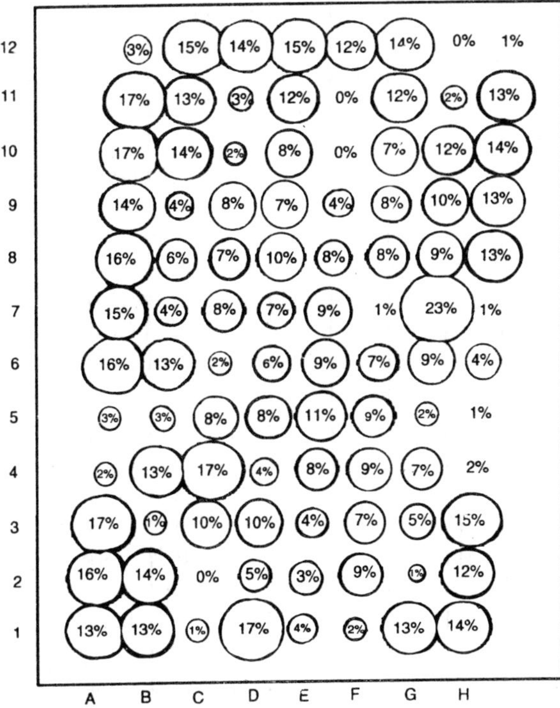


*Figure 3.19 : Graphical representation of the results of one microtiter plate obtained in the screening for an amidase with improved activity towards amethylphenylglycine amide (expressed as % conversion).*

NCIMB 40321 towards DL-α-methyl-phenylglycine amide could be improved by approximately 500%. Also in this case the screening method used was identical to that in example, now using DL-α-*methylp-henylglycine* amide as substrate. In Figure elsewhere in this chapter a *graphical* representation of the results of a typical microtiter plate is given.

***Fast LC/MS for Highthroughput Screenin of Enzymatic Activity***

Among the techniques developed for highthroughput screening, mass spectrometry offers good potential to act as a generic screening method and has gained increased attention in the past decade. Some highthroughput MS screening methods for biocatalysts have been described, without combination with *chromatography*. For example MALDI MS methods have been developed and optimized to quantify *lowmolecularmass* substrates and products of enzymecatalyzed reactions. Methods based on *electrospray* ionization mass spectrometry (ESI MS) have been developed for the determination of conversion and *enantioselectivity* of *biocatalytic* reactions, in which $10^3$–$10^4$ biocatalysts can be screened per day. Despite the high throughput that can be realized by the above mentioned flow injection analysis (FIA) MS systems, byproducts may not be detected in this mode and screening for enantioselective (bio)catalysts necessitates the use (and consequently the synthesis) of *pseudo* enantiomers (chiral compounds differing in absolute configuration and isotopic labeling only) or *pseudo* prochiral compounds. The development of simple, reliable LC/MS interfaces, most notably ESI and *atmospheric pressure chemical ionisation* (APCI), has spurred the development and acceptance of LC/MS methods, such that, nowadays, there are many laboratories that routinely use LC/MS as the primary analytical method.

The LC system is used to fractionate a preparation and the mass spectrometer is positioned as a detector, monitoring the effluent. With regard to fast LC/MS methods for the analysis of small molecules (mol wt < 1000) in *biological matrices*, an overview of these methods has recently been published. In this review article, fast LC/MS methods for analytes in, for example, *blood*, *plasma*, *serum*, *urine*, and *saliva*, are described. Both gradient and *isocratic* methods are discussed, whereby '*fast*' is defined as exhibiting a cycle time of less than 5 min. For the determination of enantioselectivity, fast chiral HPLC/ MS methods are reported using *commercial chiral* stationary phases (CSP). Despite the fact that numerous LC/MS applications for blood and plasma are described in the literature, relatively few studies deal with LC/MS applications for medium to highthroughput screening of *biocatalysts*.

Because of its sensitivity, specificity, and medium to high throughput capability, LC/MS is often the method of choice for screening unpurified *biotransformations* where low conversion numbers are expected. The throughput required with a large "*random*" screening population dictates for both the HPLC and mass spectrometer a specific setup. Fast *chromatographic* run times per sample are required, which necessitate the use of short column lengths, accompanied by sufficient efficiency. The latter may be realized by using either particulate column packings with small particles (for instance 2 or 3 lm) or silica rods. With respect to the mass *spectrometric* part, both LC/MS and LC/MS/MS can be run in both positive and negative ion modes.

*Example: Screening of a Bacterial Expression Library for Amidasecontaining Clones*

This LC/MS biocatalyst screening was set up in our laboratory for validation of the hits found with the flowinjection NMR screening assay. The screening assay was intended to determine amidase activity towards *amethyl leucine amide* as example for a,adisubstituted aamino acid amides, so the product of this reaction was *amethyl leucine*.

*Procedure : LC/MS Assay of Amidase Activity*

The two 96-well plates out of 120 containing hits according to the flow-injection NMR screening assay, were centrifuged for 10 min at 1000 × *g*. From the supernatant (cleared reaction mixture) in each well, 5 μL was transferred to shallow 96well microtiter plates each containing 245 μL *acetonitrile*/water (50/50) supplemented with $NH_4OH$ (0.5 weight %, pH 10.8).

The LC/MS system used for the screening of the two plates consisted of an Agilent 1100 system with a binary pump, a wellplate sampler with hand held controller and an Agilent 1100 mass selective detector (MSD). For HPLC, a 50 × 4.6 mm Speed Rod column (Merck) was used in combination with an isocratic mobile phase (1.5 weight % acetonitrile in 0.1% *formic acid* at a flowrate of 4 mL $min^{-1}$ at room temperature). This system operates well with 0.3–1 μL injections and a wholeplate injection program.

Analysis time per well is 20 s, resulting in a cycle time of 32 min for a 96-well plate. The mass spectrometer was operated in the selected ion mode (SIM) at *m/z* 146, that is the $[M+H]^+$ ion of the acid, and at *m/z* 145, that is the $[M+H]^+$ion of the acid amide.

To ensure quantitative LC/MS analysis, use of a complementary calibration procedure. The complementary calibration mixtures were prepared by mixing a solution of dlamethyl leucine amide HCl (67.5 mM) and a solution of *lamethyl leucine* and $NH_4OH$ (67.5 mM each) in increasing ratios covering a range of 0–2.54 mM *lamethyl leucine* (corresponding to 0–3.75% conversion of the *lamethyl leucine amide* substrate). The starting concentration of the *amino acid amide* in the 0% conversion calibration sample corresponds to the amount of amino acid amide used as substrate in the screening experiments. A conversion of 50% means that all of the lamino acid amide is converted into lamino acid and ammonia. As solvent for both solutions, the same rich medium as applied in the screening protocol was used in which *E. coli* had been cultured over 24 h at 20°C. After removing the cells by centrifugation, the supernatant was diluted 4 times with MilliQ water before dissolving the calibration curve components. The *chromatograms* obtained for the calibration solutions by SIM.

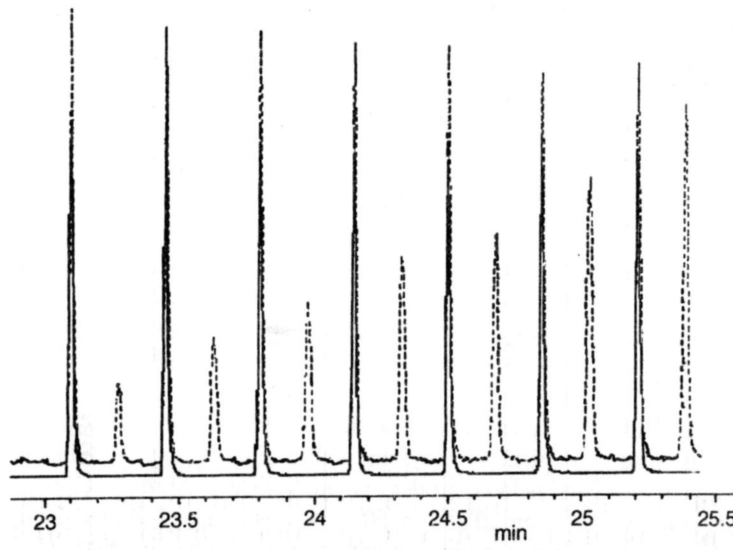


*Figure 3.20 : Chromatograms of complementary calibration mixtures recorded in SIM mode (m/z 145 is $[M+H]^+$ of amethyl leucine amide (black line); m/z 146 is $[M+H]^+$ of amethyl leucine (dashed line)). The abundance plots are normalized to full scale.*

In all calibrations carried out during the LC/MS screening assay, linear regression analysis indicated that the correlation coefficient of each single calibration curve was better than 0.994. The coefficient of variation for seven calibrations was of the order of 5%. The detection limit of conversion that could be determined was about 0.1% conversion in the case of *amethyl leucine amide*. For quantitation, the ratio of the acid and the amide (positive ion detection) was taken. By this method of internal standardization, the effect of possible evaporation is corrected. The two hits identified by flow injection NMR were also found by LC/MS assay. The percentage conversion found by using LC/MS was somewhat lower than those measured by flowinjection NMR, however, in both assays the two positive hits could be clearly identified among the whole sample population.

**Table 3.2 : Amidase hit identification (expressed as % conversion of DL-α-methyl leucine amide) in a bacterial gene library by flowinjection analysis NMR (FIA NMR) and LC/MS.**

| Plate number | Well number (% conv.) | FIA NMR (% conv.) | LC/MS |
|---|---|---|---|
| 77 | E4 | 3.7 | 3.3 |
| 97 | B5 | 3.4 | 2.2 |

*Example: Screening of Enzymatic Racemase Activity*

As a second example we describe the use of LC/MS for the screening of enzymatic racemase activity. In this specific case, clones from an expression library were screened for a *racemase* for α-amino-ε-caprolactam (ACL) using a fast *chiral* LC separation method combined with ionspray MS as the detection technique. The determination

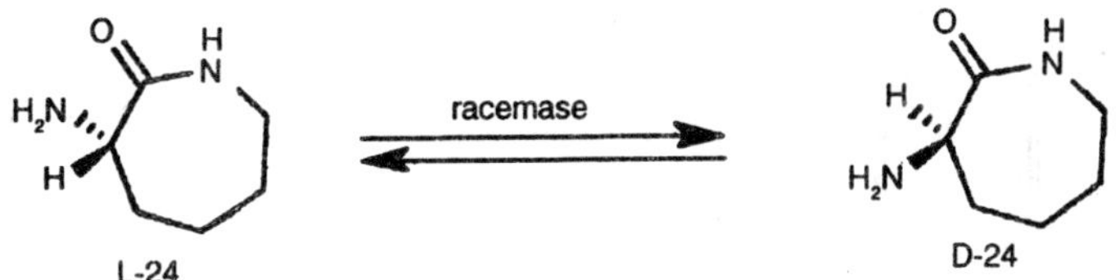


*Figure 3.21 : Reaction catalyzed by an aaminoecaprolactam (ACL) racemase.*

of the conversion in a racemase reaction necessitates the use of a method to distinguish both enantiomers of the substance, as one enantiomer serves as substrate for the enzyme, while the other is the product of the reaction.

*Procedure : LC/MS Assay for Racemase*

Using a Crownpak CR (+) column and an aqueous perchloric acid solution as mobile phase, baseline separation of the ACL *enantiomers* was obtained. Since *perchloric acid* is a nonvolatile acid, this type of mobile phase is not suited for *interfacing* with a *mass spectrometer*.

To circumvent this problem, the mineral acid may be substituted by a shortchain *organic acid* as mobile phase additive. In the *literature*, the use of *trifluoroacetic acid* and *acetic acid* as mobile phases for Crownpak columns has been reported. In our case it is necessary to protonate the ACL *enantiomers* in order to obtain sufficient retention and resolution on the crownether stationary phase. As the $pK_a$ of ACL is 7.8, an acidic pH is necessary to transform the enantiomers into their *ammonium* form. A 0.1% *aqueous* solution of formic acid (measured pH 2.6) was chosen as *eluent* instead of the nonvolatile *perchloric acid* solution. The *enantioselectivity* and resolution found by using the *formic acid* eluent were comparable with those of the *perchloric acid* mobile phase. By setting the flowrate to 1.5 mL $min^{-1}$, a run time of 1 min per sample was obtained. Using overlapping injections, a total analysis time of 56 min per 96well *microtiter plate* could be achieved. With respect to the LC/MS interface, the suitability of both the turbo ion spray and the heated nebulizer interface for the ionization of ACL were evaluated. It turned out that the sensitivity for ACL was of the same order of magnitude for both interfaces. However, in the heated *nebuliser* interface severe peak broadening occurred, which resulted in overlapping peaks of the enantiomers at a flowrate of 1.5 mL $min^{-1}$. As peak broadening in the heated nebulizer interface is flowdependent, flow splitting was tried. At lower flow rates the peak *broadening* improved, but sensitivity was lost as this type of interface is a masssensitive interface.

It was noticed that peak broadening in the turbo ionspray interface was minimal and that the peak separation obtained by LC/UV could be maintained in this MS interface. *Ionspray* MS in the positive ion mode was therefore chosen as detection technique for measurement of the $[M+H]^+$ ion of the ACL enantiomers at *m/z* 129. The D-ACL enantiomer showed hardly any retention on the column and coeluted with components arising from the biological matrix. This coelution caused some ionization suppression of the signal of the D-ACL enantiomer. The *m/z* 129 signal showed no contribution from the *biological matrix*. Selected ion *chromatograms* of racemic ACL with and without a biological matrix.

The signal for the denantiomer was suppressed by about 20%. On the other hand, the lenantiomer was hardly influenced by the biological matrix. As the goal of the assay was the measurement of the product of the racemase reaction, that is the D-ACL enantiomer, the difference in suppression between the two enantiomers was not a

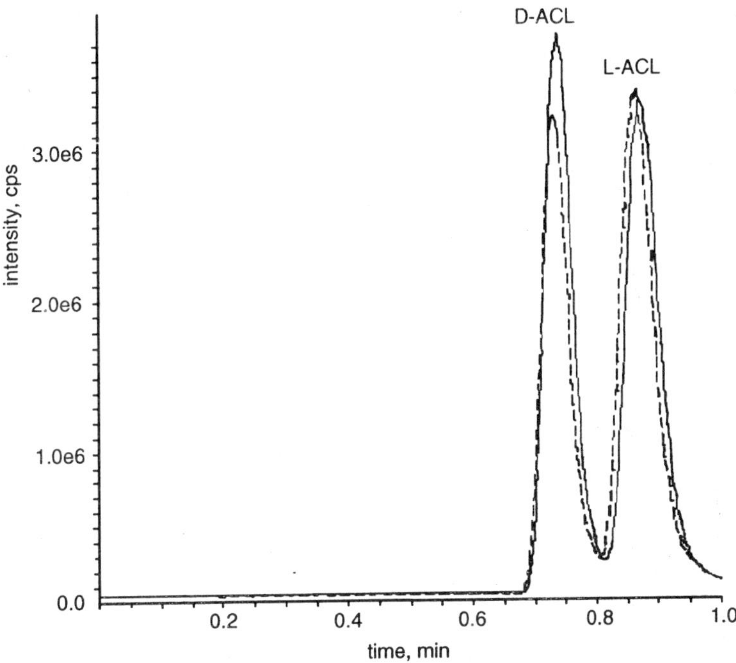


*Figure 3.22 : Selected ion chromatograms (m/z 129) showing the influence of the biological matrix on the signal intensity of the both ACL enantiomers.*

problem. In the preparation of the calibration standards, a racemase reaction was mimicked in a biological matrix that was comparable with the matrix of the expression library samples. The concentration of the calibration samples covered the range of 0–50% conversion of L-ACL, with 50% conversion implying the presence of equal amounts of D-ACL and L-ACL. Figure elsewhere in this chapter shows a *typical intraday calibration graph*, which was measured in *quadruplicate* every day.

The nonlinearity of the calibration curve is caused by the limited *dynamic range* of *ionspray ionisation*. Ionization suppression and ion signal saturation account for this effect. For the calibrated concentration range, the intraday coefficient of variation (four replicate injections of the calibration samples) was typically around 8%. The interday coefficient of variation was about 15%. The sensitivity of the method allowed a *theoretical* detection limit of 0.1% conversion. In practice this level could not be attained as the substrate of the reaction already contained 0.5% D-ACL. Because of this impurity of the substrate, the detection limit for the actual screening was set to 1% conversion.

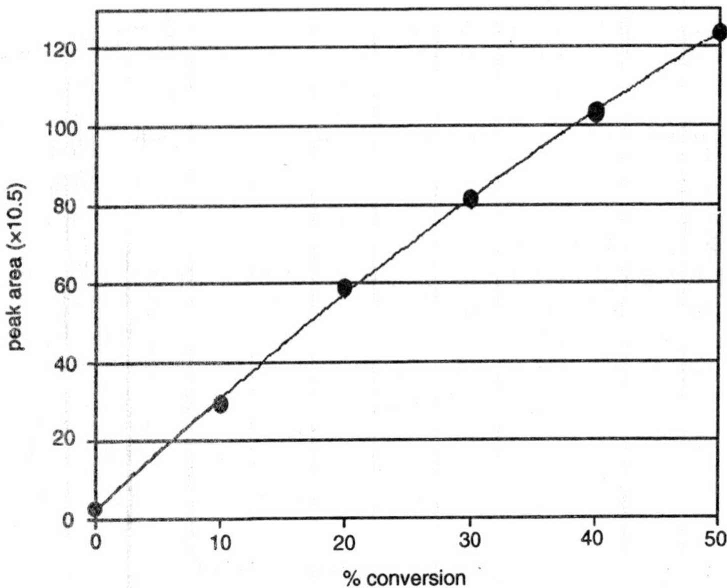


*Figure 3.24 : Typical intraday calibration graph for D-ACL, showing four replicate data points per calibration level.*

In total almost 12,000 clones from an expression library were analyzed on a single Crownpak column. No significant change in column performance was noticed after screening the total expression library.

## CONCLUSIONS

Although timelines in chemical custom manufacturing are extremely short, the two approaches outlined in this chapter will enable ontime identification of new or improved biocatalysts. The first approach requires proactive development of assay methods that are normally not suitable for this demanding business field because of their long development time. This situation is typical for reader assays requiring setup of chemical derivatization or enzymatic cascade reactions. Such chemical or enzymatic followup reactions are needed, as it is essential to assay the truly desired industrial conversion instead of using artificial colorimetric or *fluorometric substrates*. The second approach is based on use of instrumental assay techniques such as NMR and MS. As these kind of techniques are very broadly applicable ("*generic*"), leading to rather short development times, the CCM timelines can be met without any proactive development. We are convinced that by these two screening approaches one of the major hurdles for further *implementation* of *biocatalysis* in CCM, in many cases offering the cheapest and most *environmentally* benign solution, can be cleared.

# 4

# CHEMICAL COMPLEMENTATION

Directed evolution of enzymes designed *de novo* and from existing enzyme scaf-folds has the potential to produce catalysts for many useful chemical reactions. Efforts in directed evolution to date have produced a few cases in which significantly increased enzyme activity, increased *thermostability*, or modification of substrate specificity have been obtained. Notably, many of these successful examples of directed evolution have used complementation strategies as the basis for selection of improved variants, in which the function of a protein is linked to cell survival. While this approach has the advantage of being a selection rather than a screen, a major limitation is that it can only be applied to reactions naturally carried out by the cell that are inherently selectable. In order to exploit complementation to evolve enzymes for a broad range of chemical reactions, general *assays* must be developed.

Recently, our laboratory reported a complementation assay for enzyme catalysis that is reaction-independent. This assay has been called "*chemical complementation*" because it combines the power of genetic selection with the diversity of synthetic chemistry. Here, enzyme catalysis is linked to transcription of an essential gene through a small molecule yeast three-hybrid system, in which the reaction chemistry is specified in the linker of a small molecule chemical inducer of dimerization (CID). In addition to serving as the substrate for the enzyme, the CID dimerizes a DNA-binding domain and transcription activation domain to reconstitute an artificial transcription factor and activate expression of a reporter gene. Since the chemistry of the linker (i.e. the substrate) can change while the transcription readout remains constant, the assay has the advantage of being reaction-independent. This system has the potential to assay for *numerous enzymatic* activities that had previously been inaccessible to selection methods. To demonstrate the flexibility of this assay, we have applied it to such diverse reactions as b-lactam bond cleavage and glycosidic bond formation.

In this chapter, we first present a brief overview of the historical context of complementation methods, followed by an outline of the development of chemical complementation and its adaptation to detect enzyme activity. Finally, two applications of chemical compleme-ntation are described: the study of enzyme inhibitor interactions and directed evolution of a *glycosynthase*.

## COMPLEMENTATION ASSAYS

The most widely used genetic method for determination of protein function is the complementation assay. This strategy consists of modifying a cell to knock-out an essential protein and then rescuing it from death by introducing a compensatory protein on a plasmid. In the context of genetics, the protein can be derived from the genomic DNA of another organism, as in the case of gene cloning. The power of this assay lies in its ability to select from a library of thousands or millions of variants only those proteins that possess the desired function. Since the generation of variability in DNA and protein sequences is now routine, researchers have begun to apply complementation to enzymology and directed evolution. The only caveat is that a suitable selection system must be devised that can adequately distinguish functional and nonfunctional proteins, and that can also be tuned to different levels of *stringency*, depending on the desired outcome. The major limitation to this method has been the reliance on naturally selectable or screenable phenotypes. Despite these parameters, however, many promising results have emerged from complementation appro-aches.

### Early Complementation Assays

Complementation assays have historically been the tools of geneticists. In a now classic experiment, Beadle and Tatum identified enzymes responsible for the biosynthesis of essential metabolites. They introduced random mutations throughout the chromosomal DNA of the ascomycete *Neurospora crassa* by treating the organism with X-rays. Approximately 2000 individual mutated strains were then

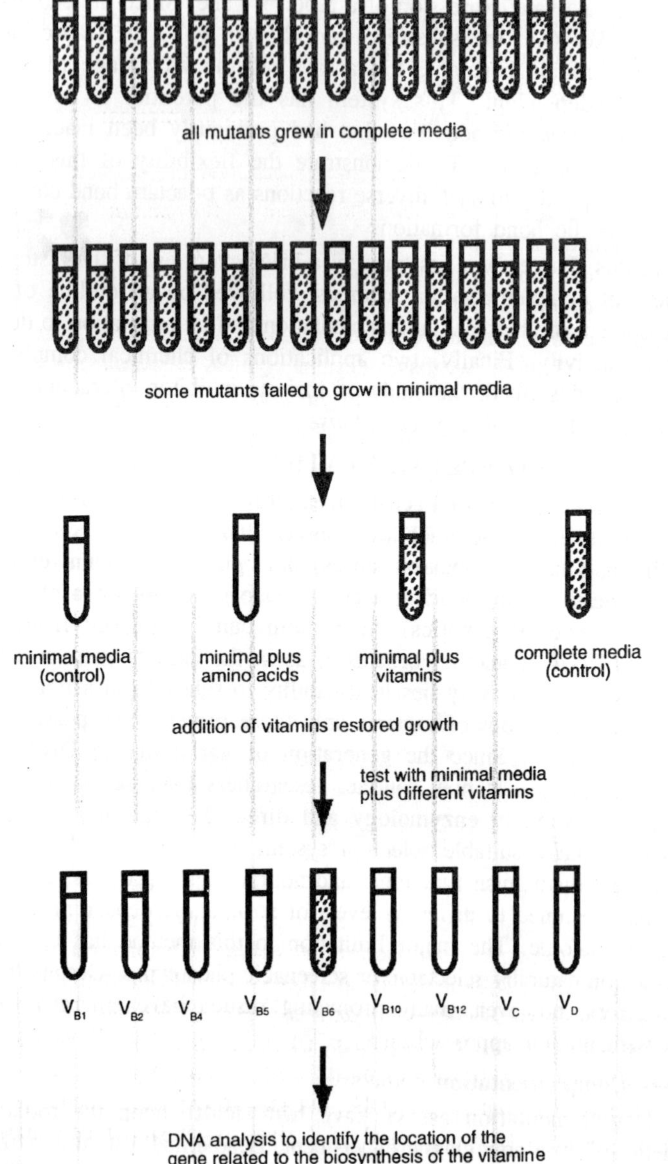


*Figure 4.1 : A genetic method to identify genes responsible for the biosynthesis of essential metabolites. Mutant ascomycetes generated by X-ray irradiation were identified by failure to grow on minimal medium. Restoration of growth was conferred by the addition of a single essential metabolite (vitamin B6).*

screened for growth in both rich and minimal medium, where the minimal medium contained only nutrients, such as biotin, that *Neurospora* cannot make itself. Three mutant strains were identified that had growth rates indistinguishable from the wild-type strain in rich media, but grew poorly or not at all in minimal medium. By adding back individual *metabolites* to the minimal medium and then measuring the cell growth rates, the three strains were shown to be unable to synthesize vitamin B6, vitamin B1, and *p*-aminobenzoic acid, respectively. This basic approach of mutating the DNA encoding the proteins and then assaying for some detectable phenotype such as cell growth is the basis for genetic analysis. Most of the genes whose function is known have been identified through this type of method, and the functions of thousands of proteins have been assayed in this manner. The contribution of Beadle and Tatum was to point out that the mutant strains could be screened based on enzymatic activity, as opposed to just cell death. Their method, however, still required several years of laborious study to identify the enzymes encoded by the genes they had mutated.

In the 1950s Benzer introduced the strategy of complementation, which provided a general solution to the problem of gene identification. In genetic complementation, a pool of plasmid DNA encoding fragm-ents of wild-type chromosomal DNA is introduced into the mutant cell line and then the DNA fragment that complements the mutation is identified based on the recovery of the wild-type phenotype. In the *Neurospora* example above, plasmid DNA that encoded fragments of the wild-type *Neurospora* genome would be introduced into the mutant *Neurospora* deficient in vitamin B6 biosynthesis. Only mutant *Neurospora* with a copy of the wild-type vitamin B6 biosynthetic gene would be able to grow in the absence of vitamin B6, allowing the plasmid DNA encoding the vitamin B6 biosynthetic gene to be distinguished readily. The recent availability of several complete *genome* sequences further simplifies genetic complementation because the plasmid DNA sequence can be compared to the complete genome sequence. In addition, it has allowed several labs to construct libraries of plasmid DNA not just encoding random fragments of *chromosomal* DNA, but instead explicitly encoding every open reading frame expressed in a given organism or cell line.

**Enzymology by Complementation**

While the cloning of proteins based on function continues to be the major application of complementation assays, they have more recently been shown to be useful tools for *enzymology*. The advantage of such an approach lies in the rapid assay of millions of mutants simultaneously. The advent of modern molecular biology has facilitated the generation of large libraries of *protein variants*. Since these libraries are genetically encoded, cell-based assays that link function of the *enzyme* to cell survival can be exploited for *global analysis* of the roles of amino acids in protein structure and function.

One aspect of protein structure that has been investigated by complementation is hydrophobic core packing. The observation that substitutions in the protein core are destabilizing led to the theory that optimal packing of side-chains is required for protein folding. While this hypothesis can be derived from first principles, it is difficult to test experimentally. However, Lim and Sauer addressed this problem by using a complementation assay to select properly folded lambda repressor proteins from a collection of mutants with all possible combinations of the core residues. They first generated a library in which seven hydrophobic amino acids in the protein core were randomized by cassette mutagenesis. Then functional repressors were selected from the library based on their ability to prevent infection of *Escherichia coli* by a repressor-defective phage. Substitutions that resulted in tightly packed cores, with hydrophobicities similar to that of the wild-type, produced active repressors. Since this experiment examined every possible set of core residues, the authors were able to draw general conclusions about the rules and patterns of protein core packing.

Their findings indicate that hydrophobicity, rather than steric constraints, play the most important role in determining core stability. This example illustrates an experimental question that could not be addressed by site-directed mutagenesis, but was well suited to a *combinatorial* mutagenesis/selection approach.

Another question of interest to enzymologists is how electrostatic interactions, *hydrogen bonding*, and conformational constraint contribute to catalysis. Hilvert and coworkers used a complementation approach to assess the role of an active site arginine (R90) in *chorismate mutase*. This enzyme catalyzes the first committed step in the synthesis of aromatic amino acids. They constructed a mutant *E. coli* strain whose *chorismate mutase* gene was inactive, and showed that the introduction of the active gene on a plasmid restored growth on a medium lacking *phenylalanine* and *tyrosine*. When all possible mutants of R90 were assayed by complementation, only the enzyme containing arginine was selected as active. This result confirmed the requirement of positive charge in that region of the active site. To assess the conformational constraint on this charge, a library was constructed in which both R90 and neighboring *cysteine* residue (C88) were completely randomized. The selection from this library produced several double mutants that were active.

**Table 4.1 : Kinetic parameters for selected chorismate mutase mutants.**

| Chorismate mutase | $k_{cat}(s^{-1})$ | $K_m$(1 M) | $k_{cat}/K_m$ $(M^{-1}s^{-1})$ | $k_{cat}/k_{uncat}$ | $k_{cat}/K_m$ $k_{uncat}(M^{-1})$ |
|---|---|---|---|---|---|
| WT | 46 | 67 | 690 000 | $4.0 \times 10^6$ | $6.0 \times 10^{10}$ |
| R90G | 0.00027 | 150 | 1.8 | $2.3 \times 10^1$ | $1.6 \times 10^5$ |
| R90K | – | – | 31 | – | $2.7 \times 10^6$ |
| C88S/R90K | 0.29 | 4300 | 67 | $2.5 \times 10^4$ | $5.8 \times 10^6$ |
| C88K/R90S | 0.32 | 1900 | 170 | $2.8 \times 10^4$ | $1.5 \times 10^7$ |

The R90L mutation was shown to be active when paired with a compensatory mutation at position 88 to serine, alanine, or glycine. Surprisingly, a lysine at position 88 could restore function when position 90 was changed to serine, showing that the placement of the cation is somewhat *flexible*. *Kinetic analysis* of these mutants showed that the removal of the cation in the R90G mutant severely diminished catalytic activity, while restoration of the cation by lysine at position 88 or 90 improved this activity by 1000-fold [9]. This result demonstrated that alternative active-site architectures are not only possible, but can be accessed experimentally. Once again, the use of a selection allowed the examination of many more mutants than could be tested by traditional screening methods. Remarkably, this study and the previous one stand alone as examples of complementation applied to enzymology. Given the power of genetic selections, it is clear that this strategy is underutilized.

**Directed Evolution by Complementation**

Complementation assays have proven particularly powerful tools for directed evolution of *proteins* with new functions. In the last section, we saw examples of protein folding and activity being linked to cell survival. These kinds of assays also lend themselves well to the selection of enzymes with new or improved properties. In a directed evolution experiment, *plasmids* that can complement the cellular defect are isolated from a starting library, mutagenized, and submitted to additional rounds of selection. This process proceeds iteratively until the desired activity is attained or convergence of the protein sequence is reached. Increasingly *sophisticated* methods for generating variation in the library have been developed in order to sample a wider array of sequences. In addition, selective pressures can be applied to the cells to increase the *stringency* of the selection, and thus glean only the most active members of the mutant library. This process attempts to recapitulate Darwinian evolution in a test tube, and has demonstrated the capacity to produce enzymes with novel and useful properties.

The earliest example of directed evolution by complementation was essentially a systematic observation of artificial selection of bacteria. Hall used a *lacZ*-deletion strain of *E. coli* to select for spontaneous mutants that could grow with lactose as the sole carbon source. The mutations all fell on the opposite side of the chromosome in a single gene, which was termed *ebg* (for evolved β-galactosidase). He first identified two phenotypes that arose from the initial round of selection: class I, which grew on lactose; and class II, which grew slowly on lactose but well on lactulose. He subjected the class I mutants to selection on lactulose and obtained class IV mutants, which grew well on both substrates. Genetic analysis suggested that class IV mutants possessed a combination of the class I and II mutations. The demonstration that enzymes with new functions could be obtained through "directed evolution" by a process that mimicked natural selection was an important step forward for the field. It is remarkable that this kind of analysis was carried out in the absence of genetic engineering methods that we now take for granted.

While Hall relied on naturally occurring mutations, chemical and enzymatic methods of mutagenesis soon came into use for generating libraries of random mutants. Typically, reagents like sodium bisulfite and nitrous acid were used to chemically modify single-stranded DNA, such that misincorporations would occur when the complementary strand was synthesized, resulting in transitions or transversions. Unfortunately, the resulting libraries were often biased, producing many mutations at "hot spots" and failing to mutagenize other positions. An important step forward was the generation of statistically random DNA libraries through the use of chemically synthesized oligonucleotides. In 1990, Knowles and colleagues devised a selection system to find active *triosephosphate* isomerase (TIM) mutants from a pool of random variants. They started with a known mutant (E165D) that was 1000-fold less active than wild-type. Then they generated a library of TIM mutants by using oligonucleotide primers that had been doped with non-wild-type bases throughout the length of the sequence, such that each member of the library had an average of one or two mutations, but retained the original E165D mutation. This library was then introduced into an *E. coli* strain that lacked the TIM gene. Selections for functional enzymes were carried out on media that contained glycerol, lactose, or both. The wild-type could grow with glycerol as the sole carbon source, whereas the E165D mutant could only grow on lactose (and grew very slowly on *glycerol*). Mutants of E165D with increased activity were selected on glycerol-only media. Six "*pseudo-revertants*" that contained additional mutations were characterized and found to possess higher catalytic efficiency than the starting mutant enzyme. These new mutations were found to cluster near the active site. The best mutant, E165D/S96P, had a catalytic efficiency of $6 \times 10^4\ M^{-1}s^{-1}$, which was 19-fold higher than the E165D mutant alone. By comparison, chemical mutagenesis only identified a single mutant with improved activity (S96P).

**Table 4.2 : Evolution of branched-chain aminotransferase activity from enzyme aspartate aminotransferase.**

| Residue No. | 7 | 34 | 37 | 41 | 126 | 139 | 142 | 269 | 282 | 293 | 297 | 311 | 326 | 363 | 381 | 387 | 397 |
|---|---|---|---|---|---|---|---|---|---|---|---|---|---|---|---|---|---|
| Wild-type | E | N | I | K | K | S | N | A | R | A | N | S | M | S | A | V | M |
| AV5A-1 | V | D | M | | | G | T | | C | V | S | | K | G | V | | |
| AV5A-7、 | V | D | M | N | R | G | T | T | | V | S | G | | | | L | L |
| AV5A-7(6) | D | M | | | | G | T | | | | S | | | | | L | |

In addition to increasing catalytic activity, directed evolution has been used to change the substrate specificity of an enzyme. For example, Yano and coworkers used a complementation assay to convert an aspartate *aminotransferase* to a branched-chain aminotransferase. First, they engineered an *E.coli* strain with a knockout in the wild-type branched-chain *aminotransferase* gene and showed that this strain was not viable unless the growth medium was supplemented with Val, Ile, and Leu. The wild-type aspartate aminotransferase gene was mutated and recombined using DNA shuffling and then introduced into the *E. coli* selection strain. The authors point out that DNA shuffling, a PCR-based method, could be used both to introduce point mutations and to generate recombined gene products. Five rounds of mutagenesis and selection were carried out. At each round, $10^6$–$10^7$

mutants could be examined, and the *stringency* of the selection could be readily increased by changing the concentration of 2-oxovaline in the growth media, the expression level of the enzyme, or the incubation time. After the final round, a mutant enzyme with 13 point mutations was isolated that had a $10^5$-fold increase in $k_{cat}/K_m$ for β-branched amino acids and a 30-fold decrease in $k_{cat}/K_m$ for aspartate. The improvement in $k_{cat}/K_m$ here is impressive and likely reflects the importance of being able to modulate the stringency of a selection—first to detect poor catalysts and finally to demand high catalytic efficiency.

The authors also address the question of how hard it is to alter specificity. By testing each of the 13 mutations individually, they showed that only six point mutations are actually required for a $10^5$-fold increase in activity toward a new substrate. Interestingly, their results also showed that all six mutations are beneficial on their own—raising the possibility that each position in a protein can be randomized independently. In a related application, the Diversa Corporation has applied a mutagenesis and selection technique, termed Gene Site Saturation Mutagenesis™ (GSSM), to evolve proteins that have improved thermostability and stereoselectivity [15]. In this approach, each position in the protein is randomized independently. With library sizes approaching $10^8$ variants, it is actually possible to randomize all positions in the protein in the same experiment.

On the challenging problem of changing the substrate specificity of an enzyme, directed evolution approaches have shown great promise. For example, Schultz and colleagues have used genetic selections to generate aminoacyl-tRNA synthetases that accept unnatural amino acids as substrates. In addition, Pabo and coworkers have adapted a bacterial two-hybrid assay to evolve zinc finger variants with defined DNA-binding specificities. In our opinion, the most successful examples of directed evolution to change substrate specificity have employed complementation assays.

Since the advent of catalytic antibodies, the hope has been that complementation assays could be used to improve the efficiency of those first-generation catalysts. However, antibodies have proven difficult to express in *E. coli* and *Saccharomyces cerevisiae* and thus difficult to improve by selection in these organisms. With recent successes in the *de novo* computational design of enzymes, it seems we may be able to return to this basic approach but with better-behaved protein scaffolds. Computational design could be used to generate the first-generation catalyst, and selection via a complementation assay could then be employed to improve the activity of that catalyst. To achieve this lofty goal, however, general complementation assays, which can link enzyme catalysis to cell growth in a way that is reaction-independent, will be needed. In the next section, we describe a system based on the yeast three-hybrid assay that fulfills this requirement.

## DEVELOPMENT OF CHEMICAL COMPLEMENTATION

As we saw in the previous section, complementation assays are very powerful, but limited to a few reactions that are inherently screenable or selectable. What is needed is an assay that has the power of a complementation growth selection but where the chemistry can be adapted to a reaction of interest. Thus, several laboratories have sought to develop high-throughput assays for enzyme catalysis that are general and can be applied to a broad range of chemical reactions. Our laboratory recently reported such an assay, which uses the yeast three-hybrid assay to link enzyme catalysis to reporter gene transcription *in vivo*. We called this assay "chemical complementation" because it resembles a complementation assay in that enzyme activity is required for the cellular phenotype, but the chemistry being complemented is controlled by the chemical linker between two small-molecule ligands and thus can be readily varied. Chemical complementation detects enzyme catalysis of bond formation or cleavage reactions based on covalent coupling of two small molecule ligands *in vivo*. The heterodimeric small molecule reconstitutes a transcriptional activator, turning on the transcription of a downstream reporter gene. Bond formation is detected as activation of an essential reporter gene; bond cleavage is detected as repression of a toxic reporter gene. The assay can be run on a large library of enzymes as a growth selection where only cells containing a functional enzyme survive. A number of well-characterized reporters for both positive and negative selections, as well as screens, have been reported.

The assay can be extended to new chemistry simply by synthesizing small molecule heterodimers with different chemical linkers as the enzyme substrates. All of the other components of the readout system remain constant while the chemistry changes. This flexibility extends this assay to a wide range of reactions, with a variety of substrate specificities and regio- and stereoselectivities. A distinct advantage of this general method is that a new assay would not have to be developed for each new enzyme target.

In this section, we describe the development of chemical complementation. Since chemical complementation hinges on the integrity of the transcriptional readout system, we initially set out to devise a robust yeast three-hybrid assay. We then adapted this three-hybrid assay into a reaction-independent enzyme assay, using β-lactamase as a model enzyme. We begin by considering three-hybrid assays generally, focusing on the dexamethasone-methotrexate (Dex-Mtx) system that was developed in our laboratory. Then we present the basic chemical complementation system and consider technical improvements made to the assay since its inception.

**Three-hybrid Assay**

Obviously, the key to chemical complementation working well is the small molecule three-hybrid system. Several three-hybrid systems and CIDs had already been reported when we began designing the chemical complementation assay. These systems supported the feasibility of our strategy. Nonetheless, we decided to begin by designing a new CID pair to provide a yeast three-hybrid system that would simplify the synthesis of the small-molecule dimerizer and give a strong transcriptional readout.

***Original Yeast Three-hybrid System***

Licitra and Liu reported the first yeast three-hybrid assay. This assay was an extension of the yeast two-hybrid assay reported by Fields and Song for the high-throughput detection of protein–protein interactions. The yeast two-hybrid assay is based on the observation that eukaryotic transcriptional activators can be dissected into two functionally independent domains, a DNA-binding domain (DBD) and a transcription activation domain (AD), and that hybrid transcriptional activators can be generated by mixing and matching these two domains. It seems that the DNA-binding domain only needs to bring the activation domain into the proximity of the transcription start site, suggesting that the linkage between the DBD and the AD can be manipulated without disrupting activity. Thus, the linkage in the yeast two-hybrid assay is the non-covalent bond between the two interacting proteins. In the yeast three-hybrid assay, the linkage between the DBD and AD is a small-molecule chemical inducer of dimerization (CID). The CID consists of two ligands that are covalently linked, and that bind to their respective receptor proteins. One receptor is fused to the DBD and the other to the AD, such that when the CID binds to both receptors at the same time, the complete transcription factor is reconstituted.

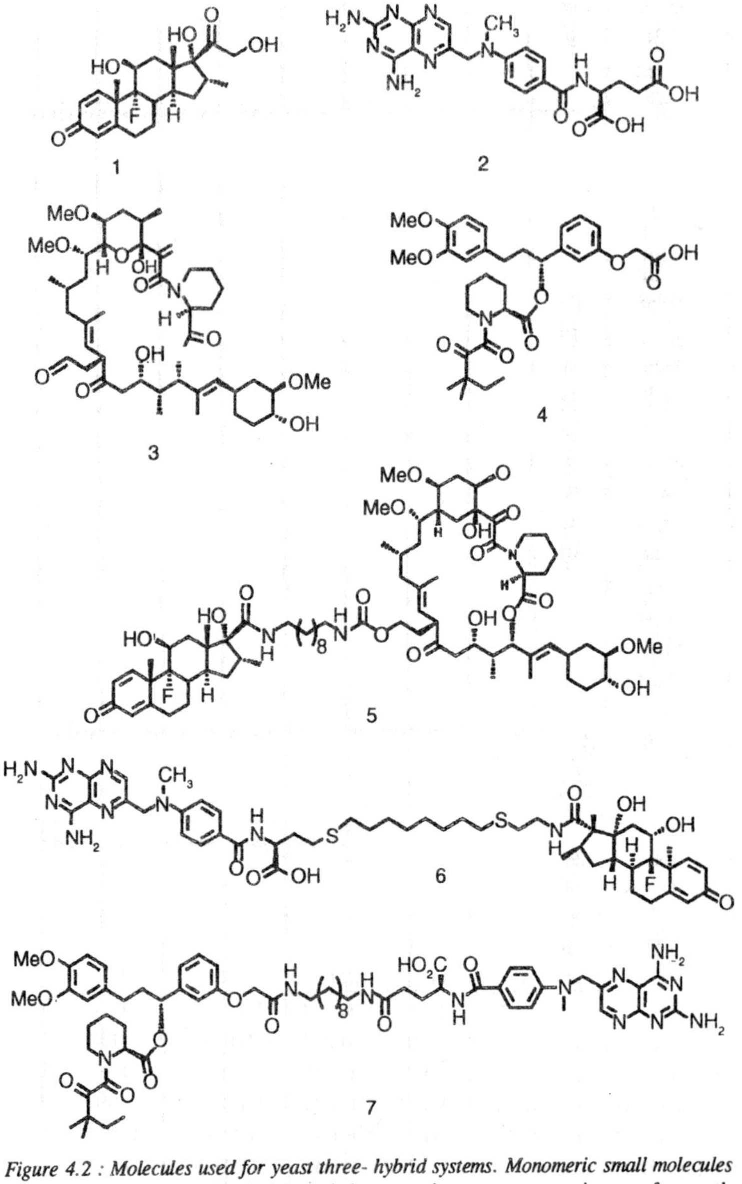

*Figure 4.2 : Molecules used for yeast three- hybrid systems. Monomeric small molecules (1–4) that bind with high affinity to their respective receptor proteins are frequently employed in yeast three-hybrid systems. Heterodimeric small molecules (5–7) that can reconstitute hybrid transcriptional activators are constructed by connecting monomers with a long, flexible linker.*

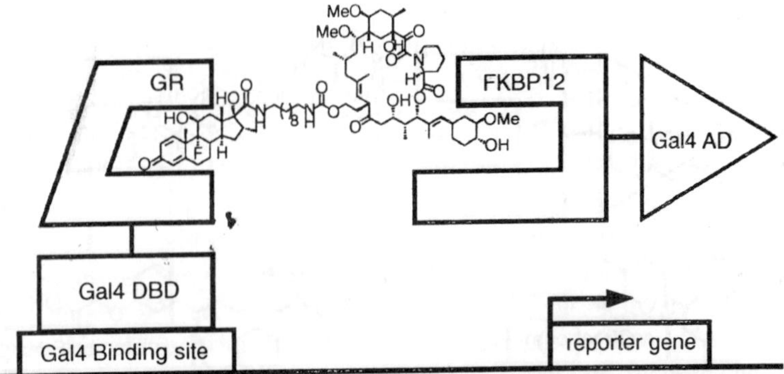


*Figure 4.3 : Dex-FK506 yeast three-hybrid system. Heterodimeric ligand (Dex-FK506) bridges a DNA-binding protein–receptor protein chimera (Gal4 DBD-GR) and a transcription activation protein–receptorprotein chimera (Gal4 AD-FKBP12), effectively reconstituting a transcriptional activator and stimulating transcription of a reporter gene. The levels of reporter transcription serve as an indicator of the efficiency of Dex-FK506-induced protein dimerization. AD = activation domain; GR = glucocorticoid receptor; DBD = DNA-binding domain.*

In Licitra and Liu's yeast three-hybrid system, the transcription factor Gal4 was split into its DBD and AD. The DBD of Gal4 was fused to the ligand-binding domain of rat glucocorticoid receptor (GR), which binds to ligand **1** (dexa-methasone, Dex) with nanomolar affinity. The AD of Gal4 was fused to yeast FK506 binding protein (FKBP12), which binds to the ligand **3** (FK506) with nanomolar affinity. Both **1** and **3** can be chemically modified without disrupting binding to their respective receptors, so heterodimer **5** (Dex-FK506) served as the CID. The reporter gene consisted of the Gal4 promoter placed upstream of the *LEU2* gene. The use of *LEU2* as a reporter gene linked reconstitution of the Gal4 transcription factor to survival of a leucine-auxotrophic yeast strain. In a proof of principle experiment, Licitra and Liu selected FKBP12 from a Jurkat cDNA library fused to the Gal4 AD. Despite its promise, there were two draw-backs to the Dex-FK506 yeast three-hybrid assay. First, polyketide **3** is a complex natural product - expensive to obtain in large quantities, requiring several steps for chemical modification, and sensitive to many standard chemical transformations. This synthetic complexity was considered a major drawback for chemical complementation, where the linker between **1** and **3** would need to be a complex substrate rather than a simple commercial linker. Second, the use of a Gal4-based system precluded the use of the regulatable *GAL* promoter for expression of the three-hybrid components or the enzyme in the eventual chemical complementation system. In our system, we sought to address these issues.

***Dexamethasone–Methotrexate Yeast Three-hybrid System***

The yeast three-hybrid assay is based on the interaction between two small molecule ligands and their respective protein receptors. In theory, any ligand–receptor pair could serve in this capacity, but certain pairs have advantages for this system. The ideal ligand–receptor pair would have a very low dissociation constant to increase the lifespan and concentration of the three-hybrid complex and thus increase the levels of reporter gene transcription. The ideal ligand would be small, membrane-permeable, and readily derivatized without disrupting its binding to the receptor. Its receptor would be monomeric and able to be expressed as a fusion protein. Methotrexate and dihydrofolate reductase (DHFR) meet these criteria perfectly. Ligand **2** (methotrexate, Mtx) is commercially available and can be synthesized readily from simple starting materials. DHFR fusion proteins have been used for a variety of biochemical applications due to the picomolar dissociation constant of **2** for DHFR. These properties made Mtx-DHFR an ideal choice for replacement of FK506-FKBP12 in the original yeast three-hybrid system. For the second pair, we chose to retain the dexamethasone–glucocorticoid receptor pair from Liu's system. The rat glucocorticoid receptor binds **1** with a $K_D$ of 5 nM, and mutants of GR with increased affinity for **1** have been isolated. The commercially available steroid **1** has been used extensively as a cell-permeable small molecule to regulate the activity and nuclear localization of GR fusion proteins *in vivo*. The heterodimer **6** (Dex-Mtx) thus served as a synthetically tractable heterodimer that could easily be modified to add the specialized linker for chemical complementation.

Another important consideration for yeast three-hybrid systems is the relative expression of the receptor–fusion proteins. The *GAL* promoter is the most tightly regulated promoter in yeast and can be induced to high levels with galactose, completely repressed by glucose, and tuned with varying ratios of galactose and glucose. Of course, this promoter could not be used in previous yeast three-hybrid systems that employed the Gal4 system because Gal4 is part of the *GAL* promoter system. Controllable expression of the protein fusions offers a number of advantages, particularly for molecular evolution. The enzyme can be expressed at high levels during early rounds of selection to obtain a low level of the desired activity. The enzyme expression levels can then be decreased during later rounds of selection to discriminate the most active enzyme variants. Inducible systems also offer a rapid means to check for false positives, by rescreening for the ability to activate reporter genes in the absence of inducer. In order to make a yeast three-hybrid system that would permit the use of the GAL promoter, we employed the orthogonal LexA-B42 system similar to that of Brent and coworkers. The *E. coli* repressor LexA served as the DBD, while the *E. coli* B42 "*acid-blob*" domain served as the AD. Neither of these components is known to interfere with the yeast regulatory systems. The selective markers on the LexA and B42 plasmids were changed from ampicillin-resistance to kanamycin-resistance in anticipation of their use in chemical complementation (*vide infra*). This modification eventually allowed the use of a different marker on the enzyme vector to facilitate its isolation following the selection. Finally, *lacZ* and *LEU2* were used for the reporter genes so that the assay could either be run as a screen or a selection.

Having made these improvements to the system, we showed that heterodimer **6** can activate *lacZ* transcription ***in vivo***. On the basis of previous studies showing that *lacZ* transcription levels correlate with the strength of protein–protein interactions in the yeast two-hybrid assay, we expected β-galactosidase activity to be a good indicator of Dex-Mtx-induced protein dimerization. Using standard β-galactosidase activity assays we determined that optimal transcription of *lacZ* occurred at concentrations of **6** greater than 1 μM. Control experiments established that *lacZ* transcription was dependent on **6**. Only background levels of β-galactosidase activity were detected when **6** was omitted. Competition of **6** with a 10-fold excess of **2** reduced Dex-Mtx-dependent *lacZ* transcription to near background levels. A ten-fold excess of ligand **1**, however, did not affect Dex-Mtx-dependent *lacZ* transcription, and higher concentrations of **1** were toxic to the yeast cells. This result may be due to differences in cell permeability between **1** and **6** or may suggest that LexA-DHFR, but not B42-GR, is the limiting reagent. In addition, when either or both receptors were deleted, only back- ground levels of *lacZ* transcription were detected. The fully assembled system produced a strong transcriptional response in the presence of **6** and provided a robust system in which to read out enzymatic activity in chemical complementation.

***Technical Considerations***

Since the introduction of the yeast two-hybrid assay in 1989, there have been a number of improvements to the basic system. Different DNA-binding and activation domains have been introduced. Convenient vectors with different bacterial drug-resistance markers, yeast origins of replication, and yeast auxotrophic markers are now commercially available. An array of common yeast genetic markers have been introduced as reporters, making it possible to test large pools of protein variants (about $10^6$) using growth selections. In addition to the simple activator system, reverse- and split-hybrid systems have been developed to detect the disruption of protein–protein interactions, and most recently a transcriptional repressor-based system was reported. These improvements to the yeast two-hybrid system are directly applicable to the three-hybrid system, and have been given detailed treatment in several recent reviews. The CIDs are an additional complexity of the yeast three-hybrid system and bring their own challenges. Two important considerations that have been examined experimentally in our laboratory are the effect of the CID linker on transcription activation and the species-dependence of the receptor proteins.

To carry out a high-throughput screen with the Dex-Mtx yeast three-hybrid assay, it proved necessary to stabilize the transcription readout. The original Dex-Mtx system showed both a high number of false positives and false negatives, and an inconsistency in the levels of small-molecule-induced transcription activation. It is generally assumed that this variability in the transcription readout comes from the use of multiple 2l plasmids. For the Dex-Mtx yeast three-hybrid system, the DBD fusion, AD fusion, and reporter gene are all on 2l plasmids, and for the chemical complementation system a fourth 2l plasmid would then be used to express the enzyme. Not only can these plasmids recombine with one another since they have large regions of sequence identity, but also the distribution of plasmid copies likely varies in each cell. The common remedy is to integrate the genes of interest into the chromosome. However, doing so decreases the number of copies of each gene from as many as 60 to one or two (haploid or diploid), which may not be sufficient for a strong transcription readout. Thus, all combinations of the DBD, AD, and reporter gene were integrated into the chromosome, and the resulting strains were tested for their levels of small-molecule-induced *lacZ* transcription. The AD and reporter gene were found to be the limiting reagents, such that the DBD and any one other component could be integrated without affecting the levels of transcription activation significantly. Use of a Dex-Mtx yeast three-hybrid strain in which the DBD and reporter gene were integrated reduced the false positive rate from 20% to 3%, a significant improvement. Interestingly, integration of the DBD actually increased the levels of transcription activation, which arguably is consistent with optimizing the concentration of DBD bound both to DNA and to the CID. For all of the applications carried out in our laboratory to date, it has proven necessary to integrate one or more components of the yeast three-hybrid system into the chromosome, and it is not simply predictable which integrated strain will be optimal for a given application.

Despite the widespread use of yeast two- and three-hybrid assays, very little is known about how transcription activation levels correlate with the strength of the binding interaction. Golemis and coworkers undertook a systematic study of the relationship between protein–protein interaction and the yeast two-hybrid readout using homo- and heterodimers for which the dissociation constants had been determined *in vitro*. From their studies they concluded that in the yeast two-hybrid system it is not possible to detect interactions with dissociation constants above 1 $\mu$M. To extend this work to the yeast three-hybrid system, we set out to examine the correlation between the strength of the binding interaction and the transcription readout using a series of mutants of a well-studied ligand–receptor pair. We chose to carry out these studies in the Mtx-DHFR yeast three-hybrid system described above. The sensitivity of the three-hybrid assay should be a function of the ligand–receptor pair being used as an anchor.

Since **2** binds to DHFR with low picomolar affinity, the use of **2** as an anchor should facilitate the detection of relatively weak interactions at the other end. To characterize the sensitivity of the Mtx yeast three-hybrid system, we considered the interaction between FKBP12 and **4** (*s*ynthetic ligand for *F*KBP12, SLF). The nanomolar dissociation constant of this pair left ample room for the design of FKBP12 mutants with reduced but measurable affinity. First, we demonstrated that **7** (Mtx-SLF) could activate *lacZ* transcription in a yeast three-hybrid system consisting of LexA-DHFR and B42-FKBP12. Then, to obtain FKBP12 variants with a wide range of binding affinities for **4**, we designed point mutants of FKBP12 on the basis of previous biochemical studies and analysis of the high-resolution crystal structure of FKBP12 with **4** and **3**. We then determined the dissociation constants for **4** and the FKBP12 mutant variants by using a fluorescence polarization assay [54]. Finally, we inserted these FKBP12 point mutants into the yeast three-hybrid assay and measured the levels of transcription activation using a *lacZ* reporter gene. This experiment showed that binding affinity in the yeast three-hybrid system correlates with transcription, but only over a dynamic range of an order of magnitude. Furthermore, this study demonstrated that the yeast three-hybrid system has a $K_D$cutoff of 50 nM when using the Mtx-DHFR anchor. Interestingly, another study, in which the Mtx-DHFR anchor is used to screen a cDNA library for drug targets, estimates the detection limit to be low micromolar. Since the readout is affected by the placement of the linker on the ligand and the folding and localization of the receptor, there may not be a simple correspondence between affinity and transcription for every interacting pair.

**Other Three-hybrid Systems**

The appeal of high-throughput screening methods to detect protein–small molecule interactions has prompted the development of yeast three-hybrid systems that use alternative strategies for transcription activation. Since any ligand–receptor pair can potentially be used in the three-hybrid system, some groups have looked to exploit increasingly high-affinity interactions as anchors. GPC Biotech introduced a drug–target interaction into a yeast three-hybrid system

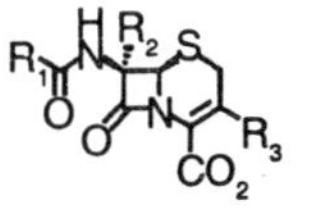

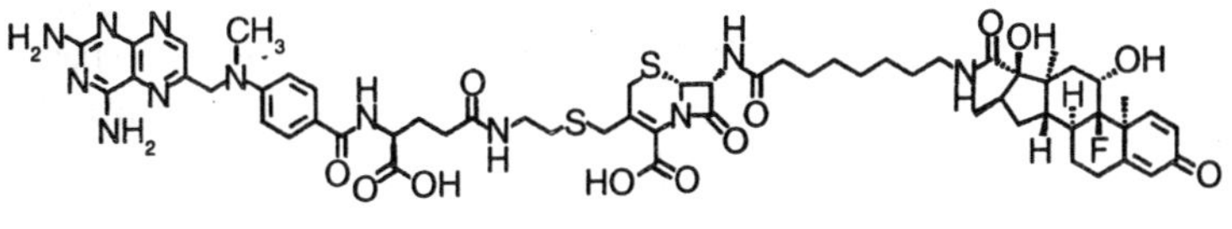

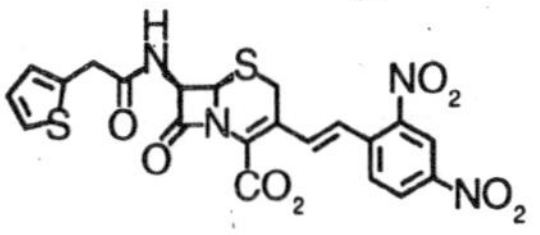

*Figure 4.4 : Molecules used in the development of chemical complementation. The cephem core β-lactam (8) is a substrate for cephalo- sporinases. Mtx-Cephem-Dex (9) (MCD) is both a substrate for cephalosporinases and a heterodimeric ligand for the Dex-Mtx yeast three-hybrid system. Nitrocefin (10) is a chromogenic cephalosporinase substrate.* o*-Nitrophenyl-β-D-galactopyranoside (11) (ONPG) and X- Gal (12) are chromogenic substrates of the b-galactosidase reporter gene (LacZ).*

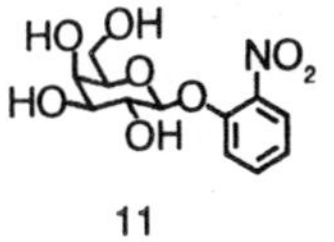

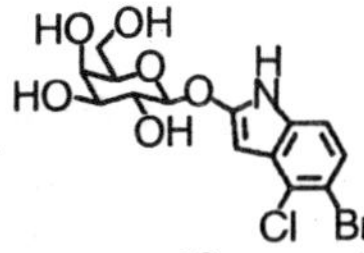

anchored by Mtx-DHFR. They then used this system to screen cDNA libraries for cyclin-dependent kinases (CDKs) that bind selectively to a given CDK inhibitor. Peterson and coworkers developed a yeast three-hybrid system anchored by the femto-molar biotin–streptavidin interaction. Taking this approach to the extreme, Johnsson and colleagues designed a system based on covalent bond formation between the CID and the DBD, using the human DNA-repair enzyme $O^6$-

alkylgua-nine-DNA alkyltransferase (hAGT) and its substrate, $O^6$-methylguanine. The DBD-hAGT fusion served as one "receptor" and DHFR-AD as the other, while the CID consisted of **2** linked to $O^6$-benzylguanine.

Another interesting possibility is the use of a small molecule to activate transcription, without the need for a protein AD. Mapp and coworkers have produced a system in which the CID consists of **2** linked to derivatives of the small molecule isoxazolidine, which interacts directly with the transcription machinery to activate reporter gene expression. Three-hybrid approaches have generated much interest as high-throughput screening tools, but the validation of these methods will require their application to practical problems of protein–small molecule interactions.

**Chemical Complementation**

To implement chemical complementation, we needed a robust small molecule yeast three-hybrid assay and an enzyme–substrate pair that could be programmed into the assay. Our optimized Dex-Mtx yeast three-hybrid system produced a strong small-molecule-dependent transcription readout. All we needed to do was introduce the enzyme on a plasmid and the substrate in the linker of **6**. Having developed the assay around a model reaction, we then established it as a high-throughput screen for enzyme activity.

***Selection Scheme and Model Reaction***

To adapt our Dex-Mtx yeast three-hybrid system to read out enzyme catalysis of bond cleavage reactions, we replaced the linkage between Mtx and Dex in **6** with the substrate for the reaction and added an enzyme as a fourth component to the system. We chose cephalosporin hydrolysis by the *Enterobacter cloacae* P99 cephalosporinase as a simple cleavage reaction to demonstrate the selection strategy. Cephalosporins are β-lactam antibiotics, and cephalosporinases are the bacterial resistance enzymes that hydrolyze and, therefore, inactivate these antibiotics. The P99 cephalosporinase is well-characterized bio-chemically and structurally and the synthesis of cephem compounds (the family described by **8**) is established. We chose to incorporate Mtx and Dex at the C3′ and C7 positions, respectively, of the cephem core ($R_3$ and $R_1$ of **8**).

Cleavage of the β-lactam bond in the cephem moiety of heterodimer **9** (Mtx-Ce-phem-Dex, MCD) results in expulsion of the leaving group at the C3′ position, effectively breaking the bond between Mtx and Dex. Thus, heterodimer **9** should reconstitute the

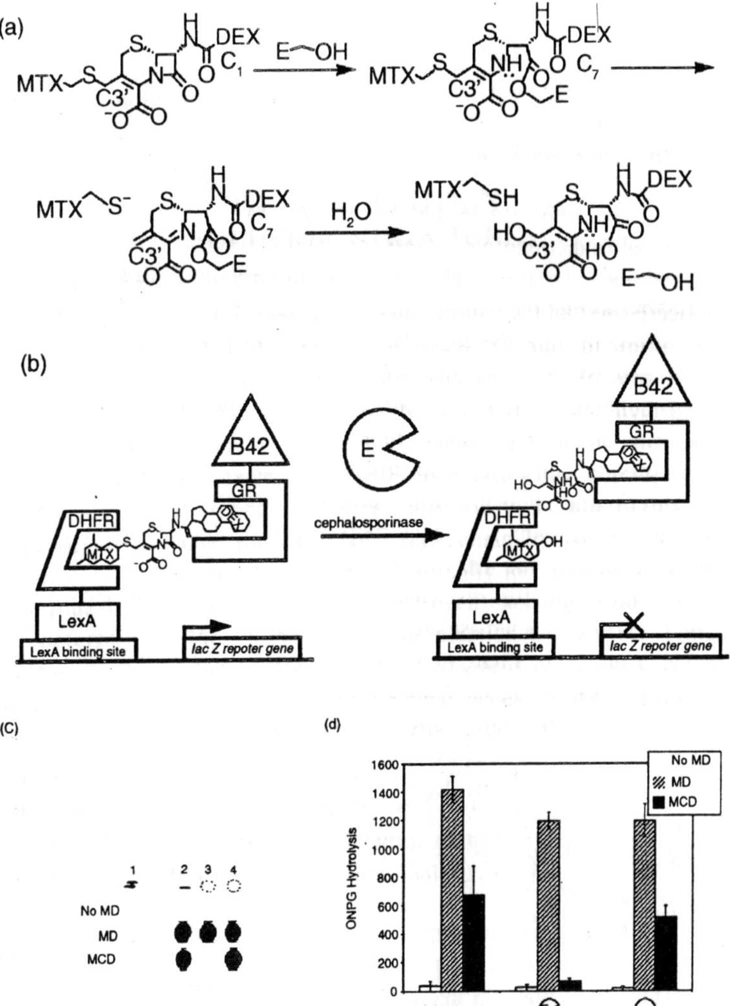


*Figure 4.5 : Chemical complementation links enzyme catalysis to reporter gene transcription. Cephalosporin hydrolysis provides a simple cleavage reaction to demonstrate the complementation strategy. (a) Cephalosporin hydrolysis by a cephalosporinase enzyme. (b) Cephalosporin hydrolysis by the cephalo-sporinase enzyme disrupts transcription of a* lacZ *reporter gene. (c) X-Gal plate assays of cephalosporinase-dependent Mtx-Cephem-Dex (MCD)-induced* lacZ *transcription. (d)* o-*Nitrophenyl-β-D-galactopyranoside (ONPG) liquid assays. Yeast strains expressing the LexA-DHFR and B42-GR fusion proteins and containing* lacZ *reporter gene and expressing no enzyme (left), P99 cephalosporinase (center), or P99 S64A cephalosporinase (right) were grown in liquid culture and assayed for β-galactosidase activity with* ***11*** *as a substrate.*

transcriptional activator, causing transcription of the *lacZ* reporter gene in the yeast three-hybrid assay. When the cephalosporinase enzyme is expressed, however, the cephem linkage should be cleaved, and protein dimerization and transcription of the reporter gene should be disrupted.

The next step was to add the enzyme to the yeast three-hybrid system and find conditions where we could observe enzyme-dependent transcription. To this end, the yeast three-hybrid strain was re-engineered so that the expression of the transcriptional activator fusion proteins and the enzyme could be regulated independently. The LexA-DHFR and B42-GR proteins were placed under the control of the fully regulatable *GAL1* promoter; the P99 cephalosporinase, under the repressible *MET* promoter. In addition, the enzyme was expressed from a plasmid containing a spectinomycin-resistance marker, whereas the transcriptional activator fusion proteins were expressed from vectors containing kanamycin-resistance markers, to facilitate isolation of the plasmid encoding the enzyme at the end of a screen or selection. Finally, no ampicillin-resistance markers were used because these markers encode b-lactamase enzymes, which, if expressed, could cleave the MCD substrate. First, we established independently that the cephalosporinase was being expressed in an active form in the yeast cells by using **10** (nitrocefin), a known chromogenic substrate for the P99 cephalosporinase. We then developed conditions where expression of the P99 cephalosporinase disrupted MCD-mediated *lacZ* transcription. These results were confirmed by using quantitative assays in liquid culture with **11** (*o*-nitrophenyl-β-d-galactopyranoside, ONPG) and demonstrated that the three-hybrid assay could be used to detect cephalosporinase activity.

***Results***

To confirm that the observed activation of the reporter gene by heterodimer **9** was truly enzyme-dependent, we carried out a number of control experiments. Using standard β-galactosidase assays both on plates and in liquid culture, we measured the levels of *lacZ* transcription in yeast strains expressing different LexA and B42 fusion proteins, enzymes, and a *lacZ* reporter gene. Transcription in each of these strains was tested in the presence of no small molecule, **6** (MD), or **9** (MCD). As can be seen, *lacZ* transcription in the strain expressing LexA-DHFR and B42-GR is small-molecule-dependent. Addition of the wild-type P99 cephalosporinase disrupts this small-molecule-induced transcription activation when the cells are grown in the presence of **9**, but not **6**. Importantly, expression of the P99 cephalosporinase has little effect on the levels of Mtx-Dex-activated *lacZ* transcription. Another important control was to show that the disruption of *lacZ* transcription was due to turnover of the cephem linkage and not merely due to sequestration of substrate **9** by the cephalosporinase enzyme. To address this question, we compared the activity of the wild-type P99 cephalosporinase in this assay with that of the inactive mutant S64A. No detectable change in the level of *lacZ* transcription was observed for cells expressing the inactive enzyme. Together, these results established that the change in transcription of the *lacZ* reporter gene is due to enzyme-catalyzed turnover of substrate **9**.

One criterion for a complementation assay is its ability to select enzymes with a desired activity from a large cDNA or mutant library. To determine the usefulness of our system for high-throughput screening, an experiment was carried out to isolate active enzymes from a large pool of inactive mutants. The yeast selection strain was transformed with a mixture of plasmids containing the inactive S64A mutant (95%) or the wild-type P99 cephalosporinase (5%). The clones were grown on medium containing **9** and **12** (X-Gal) for blue-white screening. Five blue and five white colonies were analyzed: all five blue colonies contained the expected inactive mutant enzyme, while four of five white colonies contained the wild-type active enzyme. This result indicated that the assay has a low false-positive rate, demonstrating that chemical complementation can be used reliably to screen libraries of proteins *in vivo* on the basis of catalytic activity.

### *General Considerations*

An important parameter for any enzyme assay is its ability to detect enzymes of varying catalytic activity. In an evolution experiment, for example, one might wish to select for poor catalysts in early rounds and then increase the stringency to detect better catalysts in later rounds. For chemical complementation, the key question was whether differences in transcription of the reporter correlated with the catalytic efficiency of the enzyme. To address this question, mutant P99 cephalosporinases with a range of catalytic activity were constructed and assayed. *LacZ* reporter gene activation was indeed found to correlate with catalytic activity, and the system was able to distinguish between enzymes spanning three orders of magnitude in $k_{cat}/K_m$. This dynamic range makes the assay useful for many applications, including directed evolution and enzymology. To improve this dynamic range, switching to a growth selection strategy will be necessary. In such a system, the reporter gene would be an enzyme that catalyzes the synth-esis of an essential metabolite. The advantages of this system are twofold. First, small changes in transcription are amplified into large differences in growth advantage, allowing detection of slight differences in enzymatic activity. Second, the assay can be tuned to detect enzymes in a particular catalytic range by providing more or less of the essential metabolite in the growth medium. Alternatively, changing the expression level of the enzyme could produce a similar effect. Efforts are currently underway in our laboratory to develop conditions that improve the dynamic range of the assay.

***Related Methods***

Since a general assay for enzyme activity would be an invaluable tool for drug discovery, directed evolution and proteomics, many groups have attempted to develop such a system. Some of these approaches are also based on the yeast three-hybrid, but use different selection strategies. Benkovic and colleagues have reported a three-hybrid system in bacteria in which the presence of an active enzyme results in transcription of the arabinose operon. Enzyme activity is linked to transcription by using unreacted substrate to compete with a CID for binding to the yeast three-hybrid receptor fusion proteins. In this example, scytalone dehydratase (SD) can process a scytalone analog, rendering the substrate unable to bind to the yeast three-hybrid components. Thus, when an active enzyme is present, the three-hybrid is reconstituted and transcription of the arabinose operon ensues.

In a totally different strategy, Peterson and coworkers have developed a three-hybrid-based system to detect tyrosine kinases. In this system, the peptide substrate of the kinase is covalently fused to the DBD by in-frame translation and takes the role of the CID. Recognition and phosphorylation of the peptide by an active kinase converts the peptide into a ligand recognized by an SH3 domain, which acts as a receptor. Further, this group has reported a system that makes transcription dependent on the enzymatic biotinylation of a protein substrate. Here, the CID consists simply of biotin, which the enzyme BirA covalently links to a peptide fused to the AD. The biotinylated peptide then binds to the DBD–streptavidin fusion to reconstitute the transcriptional activator.

Additional methods that use small molecules to control transcription have been put forth, including the "chemical complementation" assay of Doyle and coworkers. Kodadek and colleagues have developed a complementation system for selecting peptides that bind to proteins of interest using a negative selection, where binding of a peptide to its target protein creates a transcriptional repressor. While these methods have not been applied to enzyme catalysis, they point to a growing interest in the field of high-throughput assays that are based on complementation strategies.

## APPLICATIONS OF CHEMICAL COMPLEMENTATION

To explore the flexibility and power of chemical complementation, we have used the system to study inhibitor interactions with a β-lactamase and to improve the catalytic activity of a glycosynthase. In the first example, the system was used to screen a library of mutant β-lactamases for their interaction with third-generation cephalosporin antibiotics, in order to dissect residues that contribute to antibiotic resistance. In the second example, we added carbohydrate chemistry to the repertoire of this assay and carried out a selection for glycosynthase enzymes with improved catalytic function. In this section, we will present the results of these two applications, which exemplify the ease with which this system can be adapted to new chemistry and used to answer complex biochemical questions.

### Enzyme–Inhibitor Interactions

Although chemical complementation was designed to read out the activity of an enzyme, a logical extension of this system would be to identify or evaluate inhibitors of an enzyme. This approach is particularly attractive as a means of drug discovery using a library of small molecules to screen for an enzyme inhibitor. Alternatively, the effect of a single inhibitor on the function of an enzyme could be probed systematically by screening it against a series of mutant enzymes. In contrast to the classical approach of generating point mutants and correlating them with altered binding, chemical complementation can rapidly screen hundreds of thousands of targeted or random mutants in a functional assay to qualitatively assess the effectiveness of an inhibitor.

#### *Rationale*

β-Lactam antibiotics are the largest class of antibacterial agents administered worldwide, and the continuing development of resistance by bacterial strains to these drugs is a major public health concern. Resistance is most often mediated by the expression of β-lactamases and cephalosporinases, which inactivate the antibiotics by hydrolyzing the β-lactam bond. The β-lactamase reaction proceeds through a covalent acyl-enzyme intermediate, which in the case of resistant enzymes is rapidly hydrolyzed to regenerate the catalyst. Therefore, significant resources have been devoted to the development of highly functionalized β-lactam antibiotics that can trap these enzymes in the acyl-enzyme state, thus making them unavailable for β-lactam hydrolysis. Mutant b-lactamases that confer resistance to third-generation cephalosporins such as **13** (cefotaxime) and **15** (ceftazidime) have recently been isolated. Structural studies of these mutants suggest that residues contacting the C7 substituent of the cephalosporin are important determinants of resistance. In particular, mutating tyrosine 221 (Y221) of *E. coli* cephalosporinase AmpC to alanine or glycine appears to remove a steric constraint on the acyl-enzyme intermediate, allowing it to adopt a favorable conformation for deacylation. Additionally, biochemical evidence has shown that mutations at glutamic acid 217 (E217) and alanine 220 (A220) may also contribute to cephalosporin resistance.

In order to understand the molecular basis of antibiotic resistance, we sought to use chemical complementation to measure the effect of inhibitors on β-lacta-mases containing mutations at positions that contact the C7-substitutuent. Since genetic assays for β-lactamase activity already exist, we also used this experiment as a way to evaluate the precision and accuracy of chemical complementation for high-throughput screening.

First, we made a library of enzymes in which position 221 was mutated to all 20 amino acids and compared the inhibition profiles obtained by chemical complementation with the *in vitro* activity of these mutants. Subsequently, we used chemical complementation to screen a library in which E217, A220, and Y221 were all randomized for enzymes that were active against third-generation cephalosporins.

### *Screen Strategy*

In order to determine the inhibition profiles of mutant b-lactamases, we first set out to demonstrate that the assay could distinguish active, inactive, susceptible and resistant enzymes. Since the third-generation cephalosporins **13–16** form a long-lived acyl-enzyme intermediate with the β-lactamase, it was envisioned that these inhibitors would also block cleavage heterodimer **9** by P99 cephalosporinase, resulting in an increase in transcription of the *lacZ* reporter gene. Further, mutant β-lactamases, such as the GC1 variant cephalospo-

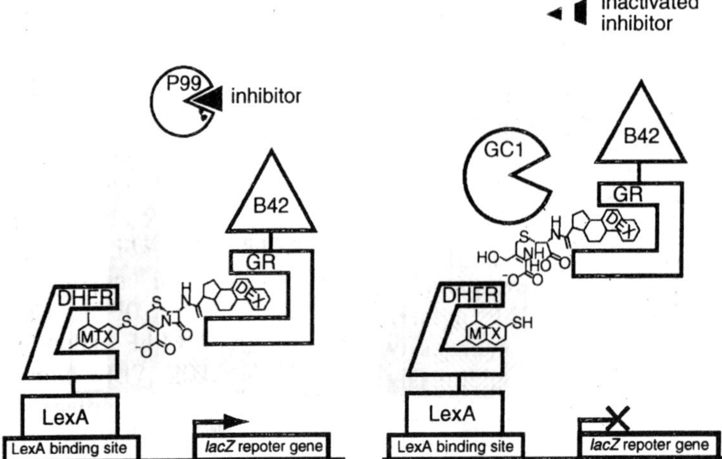


*Figure 4.6 : Chemical complementation as a screen for enzyme inhibition. The wild-type β-lactamase (left) is unable to cleave the cephem bond in heterodimer **9** due to the presence of an inhibitor bound in its active site, resulting in transcription of the* lacZ *reporter gene. Mutant β-lactamases (right) that can cleave or avoid the inhibitor can be identified using the chemical complementation assay because the enzyme cleaves the cephem bond of heterodimer **9**, turning off transcription of the* lacZ *reporter gene.*

rinase, that are not inhibited by the third-generation cephalosporins would be able to cleave heterodimer **9**, and thus could be detected based on a decrease in transcription of the *lacZ* reporter gene. This strategy should be general for enzymes that proceed through a covalent enzyme intermediate.

Initially, to establish the feasibility of this screen, we determined the ability of chemical complementation to distinguish the inhibition profiles of wild-type P99 cephalosporinase and a known resistance mutant GC1. Both of these enzymes were able to cleave heterodimer **9**, resulting in background levels of *lacZ* transcription. When the third-generation cephalosporins **13** (cefotaxime) or **14** (cefuroxime) were added to the growth medium, cells expressing the P99 cephalosporinase showed increased transcription of the *lacZ* reporter, indicating that the enzyme was being inhibited. Cells expressing the GC1 variant, however, showed relatively little change in *lacZ* transcription, indicating that this enzyme was available to cleave heterodimer **9** and thus not inhibited. By contrast, as expected, the presence of cephamycin **16** (cefoxitin) strongly inhibited both enzymes, resulting in large increases in transcription. Thus, chemical complementation could successfully classify enzymes into three

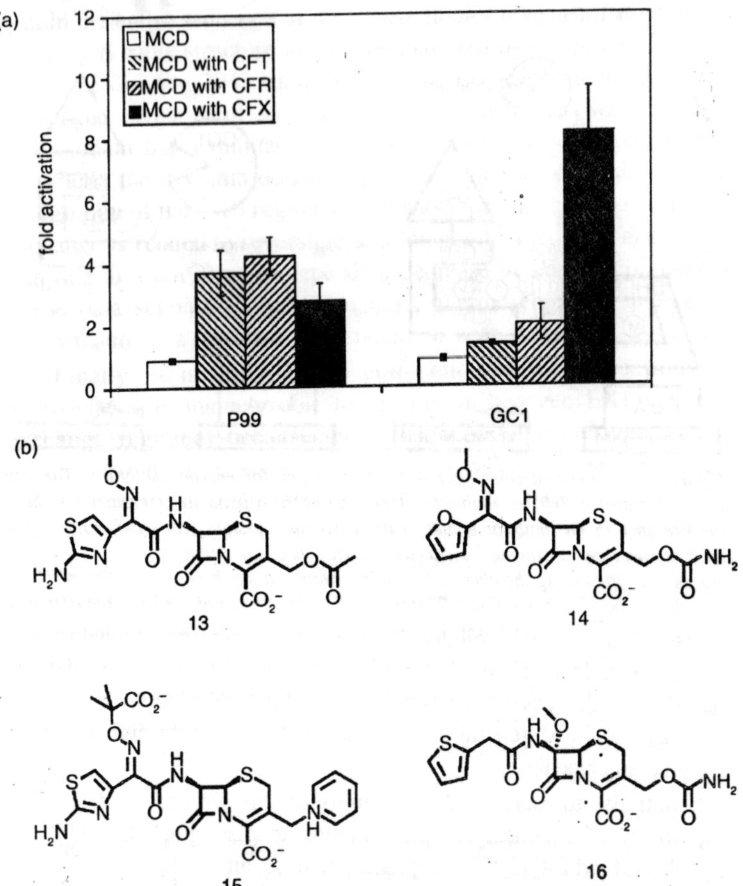


*Figure 4.7 : Inhibition profiles of two β-lacta-mases with cephalosporins. Chemical complementation can distinguish the inhibition of the P99 cephalosporinase by a series of second- and third-generation cephalosporins from that of the extended spectrum GC1 variant. (a)* o-*Nitrophenyl-β-D-galactopyranoside (ONPG) liquid assay. (b) Molecules described in this study. Cephalosporin 15 (ceftazidime, CFZ) was not tested.*

phenotypes: inactive, active and inhibited, or active and resistant. This convincingly showed that the assay could be used as an inhibitor screen.

## *Enzyme Library Screen*

Having established the feasibility of the inhibitor screen, we then carried out a pilot screen at position Y221 followed by a high-throughput screen of residues contacting the C7 substituent of the cephalosporin antibiotic. The Y221 library was characterized thoroughly to evaluate the precision and accuracy of the chemical complementation assay. The high-throughput assay then extended the assay to a large scale and explored different positions in the cephalosporin binding pocket.

For the Y221 screen, position 221 of the P99 cephalosporinase was randomized to all 20 amino acids by using degenerate oligonucleotide primers bearing the NNS codon (N = G, A, T, C; S = G, C). The library was transformed into the yeast selection strain and plated under nonselective conditions. Ninety-five randomly chosen transformants were arrayed in a microtiter plate and assayed in the chemical complementation *lacZ* screen for transcription with heterodimer **9** in the presence and absence of cephalosporin **13**. Based on this assay, each clone was assigned a phenotype: active against heterodimer **9** and resistant to **13**; active, but susceptible to inhibition by **13**;or inactive. The plasmids were isolated from each clone and sequenced in the region containing position 221 so that the phenotypes and mutations could be correlated. As expected, the Tyr and Ala mutants produced susceptible and resistant phenotypes, respectively, consistent with previous studies. Surprisingly, mutants with Phe, Leu, Val, Cys, His, Lys, Arg, and Ser all showed resistant phenotypes, in contrast to previous studies suggesting a small, hydrophobic residue was required at this position for efficient deacylation of **13**. To determine whether the screen had correctly predicted the *in vitro* activity of the mutant enzymes, a few mutants from each phenotype were expressed, purified, and assayed for activity against **13**. All of the mutants deemed resistant showed a substantial increase in their $k_{cat}$ values with **13**, compared with wild-type P99 cephalosporinase, ranging from a 430-fold increase with Y221L to over 2700-fold increase with Y221A. In addition, the $K_m$ values of these mutants were increased by about 10-fold over the wild-type enzyme, leading to an overall improvement of 100-fold in their catalytic efficiencies ($k_{cat}/K_m$). By contrast, mutants that were phenotypically susceptible to **13**, such as Y221T, showed only modest 30-fold increases in $k_{cat}$ over wild-type. Inactive mutant Y221E was substantially impaired against the heterodimer **9** substrate, with a 100-fold decrease in $k_{cat}/K_m$ compared with wild-type, despite its 400-fold increase in $k_{cat}$ for **13**. In every case where a mutant was characterized *in vitro*, the chemical complementation assay had accurately predicted the effect of the mutation. Since redundancy was built into the assay and most mutants were assayed multiple times, the precision of the assay can also be estimated as a ratio of the number of correct results to total results, 74%. This estimate of 74% is conservative because the sequencing results showed that many of the false positives/negatives

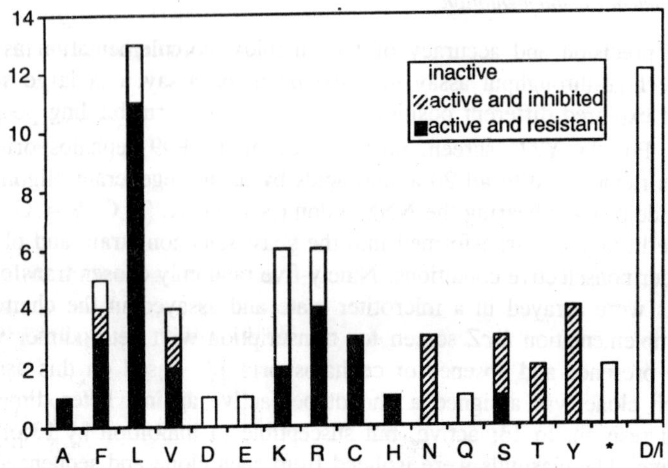


*Figure 4.8 : Y221X mutants of P99 cephalo-sporinase characterized using the chemical complementation* lacZ *screen. Mutations are grouped as hydrophobic amino acids (A, F, L, V), negatively and positively charged amino acids (D, E, K, R), polar amino acids (C, H, N, Q, S, T, Y), stop codons (*), or deletions and insertions (D/I).*

had additional mutations in the coding region. Thus, the Y221 library showed the chemical complementation assay to be robust and even revealed some surprising resistance mutations at this well-studied position.

Confident of the accuracy of the chemical complementation assay, a high-throughput screen was then undertaken to survey the effects of additional mutations in the region of the enzyme contacting the C7 substituent of the cephalosporin. A library in which positions 217, 220, and 221 of P99 were randomized using NNS codons was constructed and a random subset of 960 clones was screened with **13** in the chemical complementation assay as described above. Mutants that showed resistant profiles in this screen were characterized *in vitro* and several were found to have increased activity toward **13**. Overall, the A220N/Y221H/N226D gave both the highest $k_{cat}$ and highest $K_m$ obtained for any mutant tested, with a $k_{cat}$ that was 5600-fold over that obtained for wild-type P99 cephalosporinase and a $K_m$ that was 100-fold over that obtained for wild-type P99 cephalo-sporinase. Although the library was not screened comprehensively, an improvement in catalytic efficiency of two orders of magnitude was observed for the best mutant, D88N/D217V/A220F/Y221A. Surprisingly, a few mutants in which tyrosine was retained at position 221 also showed up to 77-fold increases in catalytic efficiency toward

**13**. This result clearly demonstrated the ability of chemical complementation to screen mutant libraries and identify improved catalysts.

***General Considerations***

The requirement of yeast cell permeability to cephalosporins led us to make two critical modifications to the system. First, the growth medium was supplemented with 1% dimethyl sulfoxide (DMSO) to slightly permeabilize the yeast cell membrane to the inhibitors. Next, the *PDR1* gene was knocked-out in the selection strain. Pdr1p positively regulates the expression of *PDR5*, *SNQ2*, and *YOR1*, which encode ABC transporters that work on a variety of small molecules. *PDR1* knockout strains have been used in a number of studies to increase the permeability of yeast to various small molecules. Together, these two modifications to the assay permitted the observation of enzyme inhibition by cephalosporins *in vivo*.

Like any high-throughput assay, chemical complementation is prone to false positives and false negatives, which result from clonal variation in the yeast three-hybrid components. Conveniently, secondary screening to identify these artifacts simply consists of multiple repetitions of the chemical complementation *lacZ* screen in the presence and absence of heterodimer **9**. Clones that fail to activate *lacZ* transcription in the presence of the CID are considered false positives for β-lactamase activity; conversely, clones that activate transcription in the absence of the CID are regarded as false negatives. In the Y221 library screen, 10 of the 95 colonies tested in this secondary screen were found to be false positives and eliminated. Since this method of secondary screening is enzyme-independent, it should be generally applicable for chemical complementation. The use of secondary screens for the assessment of yeast transcription assays is treated in greater detail elsewhere.

**Glycosynthase Evolution**

One of the major goals of enzyme research is the successful design of enzyme catalysts for any desired reaction. There are exciting recent advances in computational modeling, but the goal of designing protein sequences *de novo* that will have a specified activity is still beyond our reach. Directed evolution, on the other hand, seeks to improve catalysts by creating large libraries of protein variants and selecting for those that have improved activity. The bottleneck to enzyme discovery by this method is the design of a suitable selection system. Chemical complementation offers a way to evolve an enzyme for which there is no intrinsic selection, such as a glycosynthase.

***Rationale***

Chemical synthesis of carbohydrates is presently limited by the need for differentially protected intermediates and reactant-dependent coupling yields and stereo-control. Enzymes, with their exquisite control of both regio- and stereo-chemistry, provide a promising alternative to traditional small molecule chemistry for the synthesis of oligosaccharides. Glycosyltransferases, the natural enzymes responsible for the synthesis of oligo- and polysaccharides, and glycosidases, the enzymes that normally hydrolyze carbohydrates, have both been used for carbohydrate synthesis. The use of glycosyltransferases, however, is limited by the need for nucleotide diphosphate glycosyl donors, which are relatively more expensive and difficult to synthesize than glycosidase substrates. Oli-gosaccharide synthesis using glycosidases, not surprisingly, suffers from low yields since the enzyme also catalyzes the hydrolysis of the desired product.

Thus, alternative methods are being sought for enzyme-catalyzed carbohydrate synthesis.

Recently, Withers and coworkers demonstrated that retaining glycosidases can be engineered to glycosynthases simply by mutating the nucleophilic glutamate residue at the base of the active site to a small hydrophobic residue and using an α-glycosyl fluoride as the donor substrate. Mutation of the active-site nucleophile to a small, hydrophobic residue both accommodates the α-glycosylfluoride donor and inactivates the hydrolytic activity of the enzyme, allowing the reaction to proceed in the reverse direction. This result opened a new route for carbohydrate synthesis, and several retaining glycosidases have already been successfully converted to glycosynthases using this strategy. Directed evolution would offer an obvious route to improve the activity of these enzymes and modify their substrate specificities, except that there is no intrinsic way to screen or select for glycosynthase activity. Thus, we sought to bring chemical complementation to bear on this important class of enzymes for carbohydrate synthesis.

***Selection Scheme***

Chemical complementation is a powerful method because to look at a new reaction class, all that needs to be changed is the bond linking the two halves of the CID. For the glycosynthase reaction, we chose to model the linker after the disaccharide substrates used by the E197A mutant of Cel7B from *Humicola insolens*. Cel7B is an endoglucanase that catalyzes the hydrolysis of b-1,4-linked glucosidic bonds in cellulose with retention of stereochemistry at the anomeric center.

The E197A mutant has been shown to be a glycosynthase. We envisioned that chemical complementation could detect glycosynthase activity as formation of a bond between donor **17** (methotrexate-disaccharide-fluoride, Mtx-Lac-F) and acceptor **18** (dexamethasone-disaccharide, Dex-Cel) in a Dex-Mtx yeast three-hybrid system. For the acceptor compound **18**, cellobiose (the natural substrate of the glycosynthase) was used as the disaccharide with **1** attached at the anomeric position. For the a-glycosyl fluoride donor **17**, however, cellobiosyl fluoride could not be used, since it could also act as an acceptor and thus self-polymerize. Therefore, an epimer of cellobiose fluoride, lactosyl fluoride, was chosen that differed only in the stereochemistry at the 4' position. Ligand **2** was installed at the 6' position of the galactose unit to facilitate the chemical synthesis and because the crystal structure of Cel7B complexed with a substrate analog suggests that this position is exposed and, therefore, that the addition of **2** would not interfere with substrate binding. To validate the choice of substrates for this enzyme system, heterodimer **19** (Mtx-Lac-Cel-Dex) was synthesized *in vitro* from the donor and acceptor substrates **17** and **18** using a commercial preparation of the Cel7B : E197A glycosynthase. The complete CID activated transcription of the *LEU2* reporter gene in the Dex-Mtx three-hybrid system, demonstrating that the designed small molecules **17** and **18** were suitable for use in the glycosynthase assay.

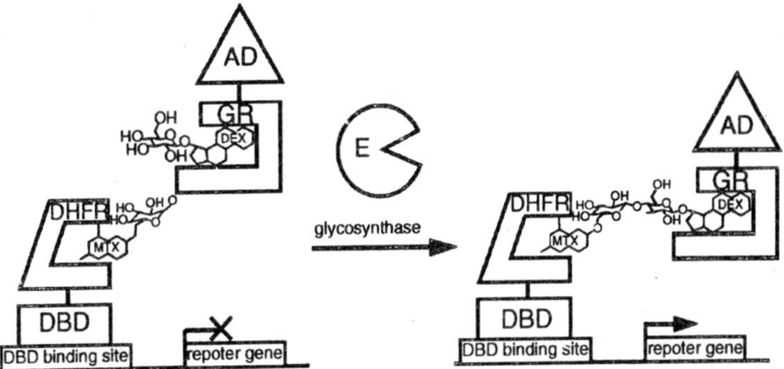


*Figure 4.9 : Chemical complementation as a high-throughput assay for glycosynthase activity. Chemical complementation detects enzyme catalysis of bond formation or cleavage reactions based on covalent coupling of two small molecule ligands.*

Once the CID for the glycosynthase was established, we could determine if chemical complementation could detect the glycosynthase activity of the Cel7B : E197A enzyme. Using standard conditions for a *LEU2* growth assay, we showed that expression of Cel7B : E197A glycosynthase in the presence of **17** and **18** conferred a growth advantage to the yeast three-hybrid selection strain, presumably because the Cel7B : E197A glycosynthase catalyzed the synthesis of heterodimer **19**. Several control experiments were carried out to confirm that the growth advantage is indeed caused by the catalytic activity of the Cel7B : E197A glycosynthase. First, we showed that transcription activation required both the donor and the acceptor substrates. The yeast three-hybrid selection strain expressing the Cel7B : E197A glycosynthase was grown in the presence of no small molecule, 10 $\mu$M **18** alone, 10 $\mu$M **17** alone, or 10 $\mu$M **17** and 10 $\mu$M **18** in media lacking the appropriate auxotrophs and leucine. Activation of the *LEU2* reporter gene results in an increase in cell growth and hence in the $OD_{600}$ for the cell culture. Cell growth was at background levels for cells grown with no small molecule or with only **18** and near background levels with only **17**. Cells grown in the presence of both **17** and **18** showed a clear growth advantage.

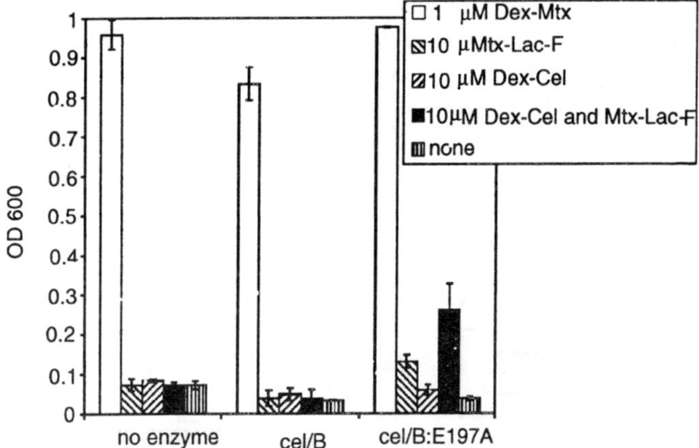


*Figure 4.10 : Chemical complementation links Cel7B : E197A glycosynthase activity to LEU2 reporter gene transcription in vivo.*

Next, we showed that transcription activation is dependent on the catalytic activity of the Cel7B : E197A glycosynthase. Cells expressing the wild-type Cel7B glycosidase, which should be able to bind the two substrates, but should be much less efficient at product synthesis, showed no growth advantage in the presence of **17** and **18**. Furthermore, cells expressing the Cel7B glycosidase were indistinguishable from cells expressing no enzyme. Together, these data suggest that the Cel7B : E197A glycosynthase activity can be detected using chemical complementation.

To demonstrate the ability of chemical complementation to carry out a highthroughput selection for glycosynthase activity, the growth advantage conferred by the Cel7B : E197A glycosynthase was used to select the glycosynthase from a pool of inactive variants. A mock library containing 100 : 1 Cel7B glycosidase to Cel7B : E197A glycosynthase was used to transform the yeast three-hybrid selection strain. Here, the Cel7B glycosidase was used as the "inactive" control. First, the transformed cells were plated under nonselective conditions to determine the transformation efficiency and ten random colonies were analyzed to establish the integrity of the library. All ten colonies were found to express the Cel7B glycosidase, as would be expected from the 100 : 1 library ratio, based on colony PCR and restriction digestion. Then the library was plated on selective media lacking the appropriate auxotrophs and leucine and containing substrates **17** and **18**. The 10 largest colonies were analyzed by colony PCR and restriction digestion. Of these 10 colonies, eight contained the Cel7B : E197A glycosynthase, and two contained the Cel7B glycosidase, which corresponds to an enrichment of 400-fold after a single round of selection. This result established that the system could select active glycosynthases from a large library and thus was suitable for directed evolution.

### *Directed Evolution*

Having demonstrated that chemical complementation could select for glycosynthase activity, we applied the *LEU2* selection to the directed evolution of Cel7B with a E197 saturation library (Figure 7.11). Position 197 was randomized to all 20 amino acids using cassette mutagenesis with an NNK codon (N = G,A,T,C; K = T,G). This library was transformed into the yeast three-hybrid selection strain and plated on selective media lacking the appropriate auxotrophs and leucine and containing substrates **17** and **18**. The 96 largest colonies from the selection plate were arrayed in microtiter plates for secondary screening to eliminate false positives. In the secondary screen, growth of the selected mutants was monitored under selective conditions with and without the two substrates. Only those colonies that showed better growth with the two substrates were considered true positives and subjected to further characterization. Of the 96 colonies picked, 35 were active in the secondary screen. The Cel7B genes from those 35 positives were PCR amplified and sequenced. Three mutants, Ala, Gly, and Ser, occurred most frequently and represent 80% of the positive clones. In addition, the Leu mutant was isolated twice and the Pro, Asp, and Thr mutants were all isolated once. Of the 10 Ala mutants isolated, three of them also had a spontaneous N196D mutation. To confirm that these mutants are indeed glycosynthases, plasmids for individual mutants were isolated, transformed back into the yeast three-hybrid selection strain, and resubjected to the secondary screen with or without small molecules. All of these variants again showed a growth advantage with substrates **17** and **18**, with the N196D/E197A variant growing faster than the E197A variant.

The E197A, E197S and N196D/E197A variants were all subsequently overexpressed in yeast and purified to allow determination of their glycosynthase activities. These three variants were assayed *in vitro* using a standard glycosynthase assay with the substrates **20** ($\alpha$-lactosyl fluoride) and **21** (*p*-nitrophenyl-$\beta$-cellobio-side). These two substrates are structurally similar to **17** and **18**, respectively. The original Cel7B : E197A glycosynthase has a specific activity of 8 $\pm$ 2 mol product $min^{-1}mol^{-1}$ enzyme. The E197S variant isolated from this selection shows a significant improvement in activity, with a specific activity of 40 $\pm$ 5 mol $min^{-1}mol^{-1}$. The N196D/E197A variant, somewhat surprisingly, is within experimental error of E197A, with a specific activity of 7 $\pm$ 1 mol $min^{-1}$ $mol^{-1}$. Thus, we were able to increase the activity of the glycosynthase five-fold in this first directed evolution experiment. We are presently expanding our efforts toward screening larger libraries through DNA shuffling, by which we expect to find mutants of significantly higher activity.

### *General Considerations*

For any high-throughput assay, a secondary screen is important for ruling out false positives. The secondary screen for glycosynthase activity simply consisted of the chemical complementation growth assay, repeated in both the presence and absence of small molecules **17** and **18**. Only those clones that exhibited small-molecule-dependent growth were considered to be true positives. Additionally, it is important to demonstrate that cell growth is enzyme-dependent and not due to random mutation in the strain. This was easily accomplished by isolating the plasmid of the putative active enzyme and retransforming it into the selection strain to repeat the assay.

The number of mutants that can be screened or selected in chemical complementation is presently limited by the transformation efficiency of yeast. While this was not a concern for the small library considered here, the largest libraries that can be considered are on the order of $10^5$–$10^7$. For the E197 saturation library, the transformation efficiency was $10^6$colonies per microgram of DNA, showing that the yeast three-hybrid selection strain is not impaired in its transformation efficiency. Yet, for larger library sizes ($10^8$–$10^{10}$), the system will have to be moved to *E. coli*, and we have made efforts in that direction.

## CONCLUSION

As we have seen, complementation approaches are a powerful means to select active enzymes from a large library of proteins. Here, we have presented chemical complementation as a reaction-independent assay that links enzyme activity to cell survival. The desired chemistry is programmed into a small molecule such that bond cleavage or bond formation reconstitutes a yeast three-hybrid transcriptional activator that drives expression of an essential reporter gene. The readout system, Mtx-Dex, LexA-DHFR, B42-GR and the reporter gene can all remain constant while the chemistry changes. Thus, all that needs to be changed for each new reaction is the Mtx-substrate-Dex, or Dex-substrate and Mtx-substrate molecules synthesized in the lab, and the enzyme library. This assay could be used to engineer glycosyltransferases, aldolases, esterases, amidases, and "diels-alderases," all with a variety of substrate specificities and regio- and stereoselectivities. By converting the assay to a coupled enzyme assay, it may even be possible to detect oxidases and reductases.

In addition to providing a powerful selection for the evolution of enzymes with new activities, there should be many uses for a reaction-independent high-throughput assay for enzyme catalysis. The assay can be used to study enzyme function, either to test hundreds of mutants to identify amino acids important for the catalytic activity of an enzyme or hundreds of different molecules to determine the substrate specificity of an enzyme. Likewise, the assay could be applied to drug discovery by screening libraries of small molecules on the basis of inhibition of enzyme activity and a change in transcription of the reporter gene. A distinct advantage is that a new assay would not have to be developed for each new enzyme target.

Finally, this assay should be particularly well suited to proteomics. A battery of Mtx-Dex substrates with different substrates as linkers could be prepared and then used to screen cDNA libraries to identify enzymes that fall into common families, such as glycosidases or aldolases. Because mammalian, as well as yeast, three-hybrid assays are standard, the assays could be carried out in the endogenous cell line, ensuring correct posttranslational modification of the proteins. The key to all of these applications is a robust assay for enzymatic activity. Chemical complementation promises to be a powerful high-throughput assay for a wide variety of enzymatic reactions.

# 5

# PROTEIN CONFORMATION

Modern mass spectrometry (MS) is well known for its exquisite sensitivity in probing the covalent structure of macromolecules, and for that reason, it has become the major tool used to identify individual proteins in proteomics studies. This use of MS is now widespread and routine. In addition to this application of MS, a handful of laboratories are developing and using a methodology by which MS can be used to probe protein conformation and dynamics. This application involves using MS to analyse amide hydrogen=deuterium (H=D) content from exchange experiments. Introduced by Linderstøm Lang in the 1950s, H=D exchange involves using $^2$H labeling to probe the rate at which protein backbone amide protons undergo chemical exchange with the protons of water. With the advent of highly sensitive electrospray ionisation (ESI)-MS, a powerful new technique for measuring H=D exchange in proteins at unprecedented sensitivity levels also became available. Although it is still not routine, over the past decade the methodology has been developed and successfully applied to study various proteins and it has contributed to an understanding of the functional dynamics of those proteins.

## INTRODUCTION

The advent of soft ESI and matrix assisted laser desorption ionisation techniques has accelerated the emergence of MS in the field of protein interaction studies. The coupling of these ionisation methods with modern mass analysers (e.g., ion traps, triple quadrupole, and time of ûight [TOF] analysers) provided breakthroughs in the biochemist's ability to more accurately determine the molecular weights of large biomolecular assemblies up to several hundred thousand daltons. Developments in microtechniques for sample preparation and handling has further advanced the role of MS for protein identification and primary structure elucidation. In particular, ESI MS has opened up opportunities for applications of MS in the fields of analytical biochemistry and structural biology largely because of its capability of analyzing proteins from aqueous solution under nearly physiological conditions of concentration, pH, and temperature, as well as its compatibility with microseparation techniques.

Of particular interest is that new applications of biomolecular MS are emerging in the field of noncovalent interactions and protein folding studies. Insight into the biological processes that occur at the molecular level and possible ways in which proteins might carry out their specific functions can be gained from a knowledge of protein folding mechanisms, their higher order structure, and the interactions of these proteins with other molecules. The question here is how can MS contribute to develop a greater understanding of the interactions in proteins that are responsible for protein stability, function, specificity, and assembly. What kind of information can be expected from MS based approaches?

Since the pioneering work of Hvidt and Linderstrom Lang, Hvidt and Nielsen, and Linderstrom Lang, hydrogen isotope exchange experiments have been used to gain information about the structural stability of protein conformers and an understanding of the mechanisms involved in protein mobility. Amide hydrogen isotope exchange studies provide information on protein folding and unfolding to and from the native state, on local fluctuations of the native state, and on ligand–protein interactions. There are essentially two competing processes during which hydrogen isotope exchange takes place. One involves internal or local fluctuations from the native state, and the other involves subglobal or global unfolding. In either case, hydrogen bonds are broken and the amide hydrogens are exposed to solvent so that exchange can occur. Methods to monitor hydrogen exchange have included tritium counting and hydrogen–deuterium (H=D) exchange in combination with spectroscopic methods such as infrared, ultraviolet, Raman resonance, and nuclear magnetic resonance (NMR). With increasing field strengths and multidimensional methods, NMR became the instrument of choice for studying hydrogen isotope exchange in proteins since the late 1970s. However, MS, particularly since the inception of ESI MS at the beginning of the 1990s, has become increasingly popular for monitoring hydrogen isotope exchange because of some decisive advantages it has over NMR. The unique advantages of MS are that it provides the ability to (a) determine hydrogen isotope content in the micromolar as opposed to the millimolar range, (b) access proteins that aggregate at high concentrations, (c) observe the hydrogen isotope content in proteins greater than 20 kDa, (d) observe coexisting conformers simultaneously, (e) work with mixtures or nonpurified samples, (f) determine the exact hydrogen isotope content and the location by analyzing peptides, and (g) detect correlated exchange, which is fundamental for understanding the exchange and folding process. The chief advantage of NMR is that the method gives the average hydrogen isotope content at many specific sites, although MS is about to play an increasingly important role here, too. The two methods yield complementary hydrogen isotope exchange information.

MS also benefits the investigation of protein conformations in other ways. The observation that charge state distributions displayed in ESI mass spectra of proteins are reflections of the net charges that the proteins carry in solution makes ESI-MS a powerful technique for detecting conformational changes in proteins. MS-based approaches in combination with H/D exchange experiments have proven to be highly efficient for probing the higher order structure of proteins, to study the conformational stability of proteins, and to determine the dynamics of conformational transitions. The aim of this contribution is to outline some approaches. In this chapter, emphasis is on probing the alterations in protein conformational stability and dynamics caused by environmental changes or chemical modifications. For other comprehensive reviews on H/D exchange in combination with MS, see also Smith et al., Yan et al., Eyles and Kaltashov, and Busenlehner and Armstrong.

## ESI-MS FOR PROBING SOLUTION-PHASE PROTEIN CONFORMATIONS

### The ESI Method

ESI-MS has emerged as a rapidly progressing technique to study biomolecular interactions due to its unique ability to directly analyse peptides, proteins, and even noncovalent complexes from solution. In the ESI process, multiply charged protein ions or charge states are generated, transferred intact into the gas phase, then introduced into the mass spectrometer and analysed according to their mass-to-charge (m/z) ratios. Because proteins have multiple protonation/deprotonation

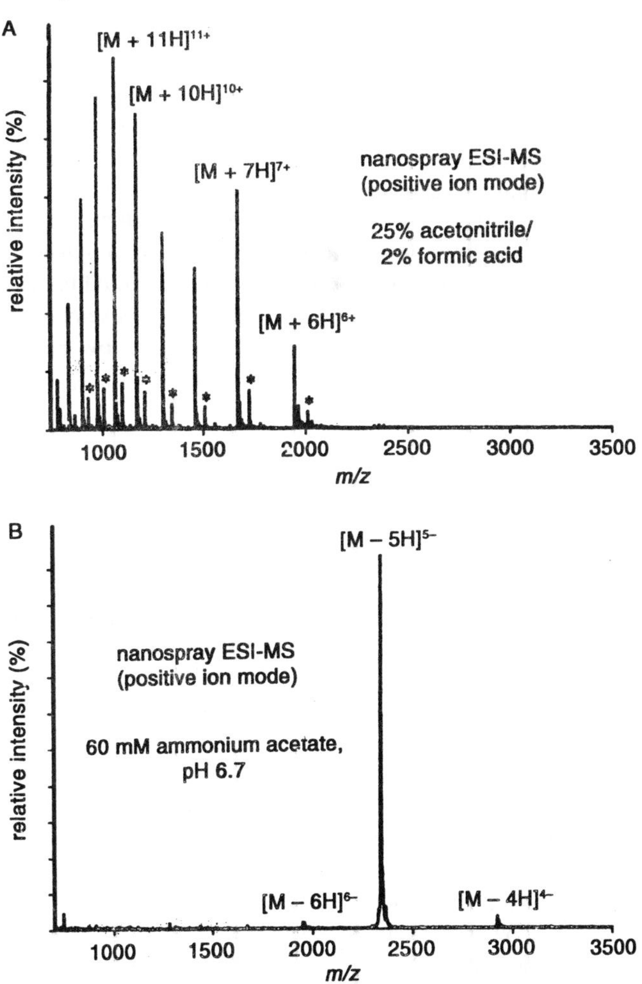


*Figure 5.1 : Nanospray electrospray ionisation (ESI) mass spectra of oxidized Escherichia coli thioredoxin (2 mM) (A) in 25% acetonitrile containing 2% formic acid measured in the positive ion mode and (B) in 50 mM of ammonium acetate, pH 6.7 measured in the negative ion mode.*

sites, a typical ESI mass spectrum of a protein M contains a series of ion peaks $(M + nH)^{n+}$ or $(M - nH)^{n-}$, in which each ion peak represents a population of protein molecules carrying n charges. The series of ion peaks, $(M + nH)^{n+}$ or $(M - nH)^{n-}$, observed for a protein under defined conditions in an ESI mass spectrum is called the charge-state distribution or charge-state envelope.

Proteins have traditionally been analysed by ESI-MS using aqueous solutions containing an organic modifier (e.g., 10–50% [v/v] methanol or acetonitrile) and organic acids (e.g., 1–5% formic or acetic acid). Under these conditions, ESI-MS allows mass determinations of proteins with high mass accuracy. However, because these conditions may cause denaturation of the protein, information regarding the conformation is lost. It is, therefore, essential to analyse proteins in solutions that favour the conformational states of interest. For instance, if the native conformation is of interest, then the protein should be analysed in a solvent that favours the native conformations, for example, mildly acidic conditions (pH 5–7), no organic solvents, and low ESI source temperature.

**Charge-State Distributions**

One of the first applications of ESI-MS to study protein conformational changes involved the analysis of charge-state distributions in ESI mass spectra. Proteins that are electrosprayed from solutions and retain their native folded states tend to have narrow charge-state distributions centered around molecular ions with low net charges, whereas denaturing solution conditions (i.e., nonnative physiological pH, high organic solvent concentrations, and elevated temperatures) or the reduction of intramolecular disulfide bonds give rise to charge-state distributions encompassing molecular ions carrying many more charges. The relationship between high charge states and the denatured state of the protein was first observed in 1990 for the acid-induced unfolding of cytochrome. Numerous proteins have since been monitored by ESI-MS under different solution conditions, supporting the general consensus that the solution-phase conformation of a protein has dramatic effects on the charge-state distribution observed in its ESI mass spectrum. Kaltashov and coworkers have described a chemometric approach to separate the contributions of the charge-state envelope of the individual conformers to the overall charge-state envelope by using singular value decomposition for extracting the number of significantly populated conformers. Validation of this method has been carried out by monitoring acid- and alcohol-induced equilibrium states of model proteins such as chymotrypsin inhibitor 2, ubiquitin, and apomyoglobin, which obey a two-state, three-state, and four-state folding model, respectively.

Still only limited information is available to fully understand the thermodynamic and kinetic processes involved in the generation of multiply charged protein molecular ions and their transfer into the gas-phase during the ESI process. Two models are discussed. In the charged residue model, the droplets generated during the ESI process shrink by solvent evaporation until only single ions remain. In the ion evaporation model, ions are desorbed from the surface of the droplet before the ultimate droplet size is reached. These two models have been refined, and at present, it is hypothesized that the direct ion emission of highly solvated molecular ions by asymmetrical droplet fission or the direct emission from the tip of a liquid cone formed under the influence of an electrical field (as in the case of nano-ESI-MS) describes best the experimental observations that (a) the charge-state distribution depends on the conformational structure present in solution and the localisation of the protonation=deprotonation sites in the macro-molecular ion and (b) the survival of intact noncovalent complexes in the gas phase. It is believed that higher degrees of protonation as observed for proteins electrosprayed from denaturing solutions can be explained, at least in part, by the more extended or less compact conformation of the denatured proteins because of an increase in surface accessibility and changes in pKa values at the protonation sites. However, one may speculate, on the basis of funnel-like protein folding landscapes, that the broader charge-state distributions for proteins electrosprayed from denaturing solutions may actually reflect the conformational heterogeneity and diversity associated with highly disordered unfolded conformational states. On the other hand, the narrow charge-state distributions observed for proteins electrosprayed from "native-like" solution conditions reflect the well-defined conformational properties of a highly compact folded state.

It should be noted that the charge-state distribution of a protein depends to some degree on external parameters associated with the intrinsic properties of the analysis technique, such as ion formation, ion transmission, and ion detection. Therefore, if one attempts to quantify changes in charge-state distributions, the external factors should be carefully controlled and evaluated. However, compared to the drastic change of the appearance of the charge-state distribution in ESI mass spectra as a result of a conformational transition, these external factors influence the appearance of the charge-state distribution only to a minor extent

## Comparison of ESI-MS with Other Spectroscopic Methods

While ESI-MS has emerged as a powerful tool for obtaining potential structural descriptions of proteins, it is highly desirable to correlate changes in charge-state distributions in response to conformational changes within a protein with data from other biophysical methods to confirm or deepen the understanding on which molecular events ESI-MS is capable of reporting and to be able to evaluate the ESI-MS–derived data critically. An example in which the charge-state distribution as observed in ESI-MS was successfully used for monitoring conformational changes in solution is the heat denaturation study of E. coli thioredoxin (TRX).

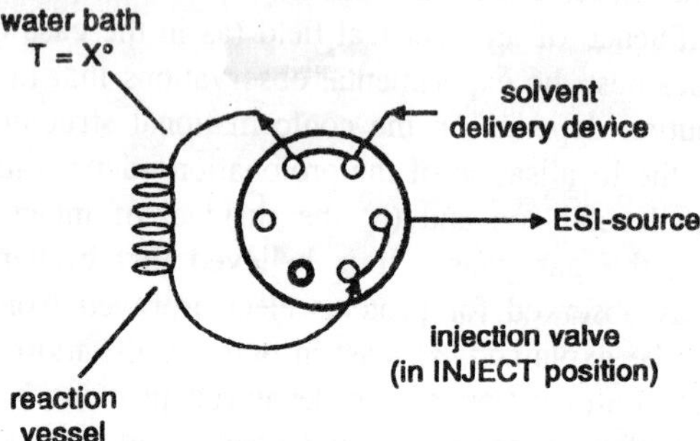


*Figure 5.2 : The protein solution is injected into the sample loop assembled in an injection valve, immersed in a water bath maintained at the desired temperature.*

For the acquisition of the temperature-dependent ESI mass spectra, a continuous flow setup was used. At ambient temperature, the acquired mass spectra of TRX exhibit a narrow charge-state distribution at higher m=z values encompassing the sixfold to ninefold protonated ions, with the $(M + 8H)^{8+}$ ion as the most abundant charge state. It is believed that these charge states represent molecular ions that were derived from the compact tightly folded conformational states of TRX. Below m=z 1200, a second series of molecular ions is observed, with low intensity representing the 10- to 15-fold protonated molecular ions, which are thought to relate to the denatured less compact conformational states of TRX.

However, the charge-state distribution of TRX equilibrated at higher temperature is dramatically different. Above 40°, the charge-state distribution at lower m=z values encompassing charges 10+ to 15+ becomes more intense, whereas the intensity of the charge states centered around $(M + 8H)^{8+}$ decreases. Above 65°, the charge-state distribution is centered around the 12-fold protonated ion peak, which then dominates the appearance of the ESI mass spectrum. Heat denaturation curves were deduced from the temperature-dependent charge-state distribution by plotting the average charge state <c> of an ESI mass spectrum or the ion peak ratio IF=(IF +IU) versus the temperature.

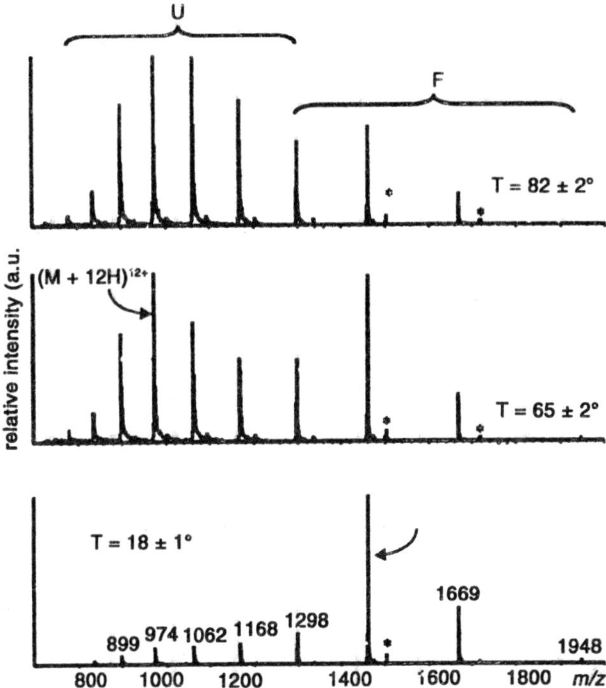


*Figure 5.3 : Electrospray ionisation (ESI) mass spectra of Escherichia coli thioredoxin at the temperatures (A) 82 ± 2°, (B) 65 ± 2°, and (C) 18 ± 1° in 2% acetic acid.*

It is important to note that at the temperatures studied, TRX showed only two distinct charge-state distributions that exhibit temperature-dependent changes in intensity without shifting of their charge-state maxima. In the case of the thermal unfolding study of E. coli TRX, the temperature-induced unfolding transition, derived from the ESI mass spectra, paralleled almost exactly with the unfolding transition observed in near ultraviolet (UV) circular dichroism (280 nm), a method that reports on changes in the environment of aromatic side chains of amino acid residues. These data together with data from H=D exchange experiments suggest (a) that the thermal unfolding of TRX under the applied conditions can be described as a two state unfolding transition allowing the estimation of Tm, the melting temperature, and LHm, the enthalpy change for unfolding and (b) that charge state distributions as observed by ESI MS reflect on the intactness of the hydrogen bonding network. These results support the emerging view that ESI-MS is a technique that is potentially capable of monitoring subtle changes in the tertiary structure.

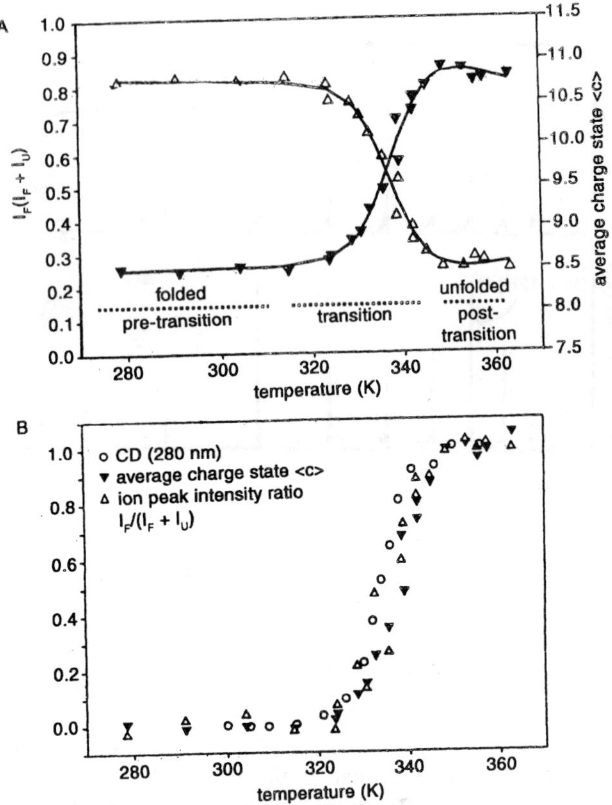


*Figure 5.4 : (A) Heat denaturation curves of Escherichia coli thioredoxin in 2 (v=v) % acetic acid deduced from the temperature dependent charge state distribution obtained by electrospray ionisation (ESI)–mass spectrometry (MS): (A) ion peak intensity ratio and (▼) average charge state <c> as a function of temperature.*

The relative error of the thermodynamic data obtained from the analyses of charge-state distributions was estimated to be in the range of 10–20%.

$$c = \frac{\sum_c I_c c}{\sum_c I_c} \qquad (1)$$

Using the same experimental approach, the global stability of oxidized *E. coli* TRX was compared to a chemically modified TRX, namely (S-(2- Cys$^{32}$ethyl)-glutathione)thioredoxin (GS-TRX). GS-TRX is a possible protein alkylation product of S-(2-chloroethyl)glutathione, a cytotoxic compound derived from glutathione and the xenobiotic

**Table 5.1 : Thermodynamic Data for the Unfolding Transition of E. Coli Oxidised Thiroedoxin Derived from Near-UV Circular Dichroism Spectroscopy and the Analyses of Charge-State Distributions as Observed by ESI-MS**

| Method | Tm (°) | AHm (kcal=mol) |
|---|---|---|
| Near-UV circular dichroism | 60 | 78 |
| ESI-MS | | |
| Ion peak intensity ratio | 64 | 74 |
| Average charge state | 65 | 54 |

1,2-dichloroethane. The heat denaturation curves deduced from the temperature-dependent charge-state distributions obtained by ESI-MS indicate that GS-TRX possesses a lower structural stability compared to oxidized TRX but a significantly higher stability compared to reduced TRX. These data together with data from other optical spectroscopic methods and H=D exchange experiments indicate that the glutathionyl moiety may interact with parts of TRX, which results in an increase in conformational stability. This hypothesis has been substantiated by energy minimisation studies on (S-(2-Cys[32]ethyl)-glutathione) thio redoxin, which propose hydrogen bonding and electrostatic interactions between the thioredoxincore and the glutathionylmoiety.

Charge-state analysis was also applied for determining the conformational consequences of metal ion binding to the colicin E9 endonuclease (E9 DN ase) by taking advantage of the unique capability of

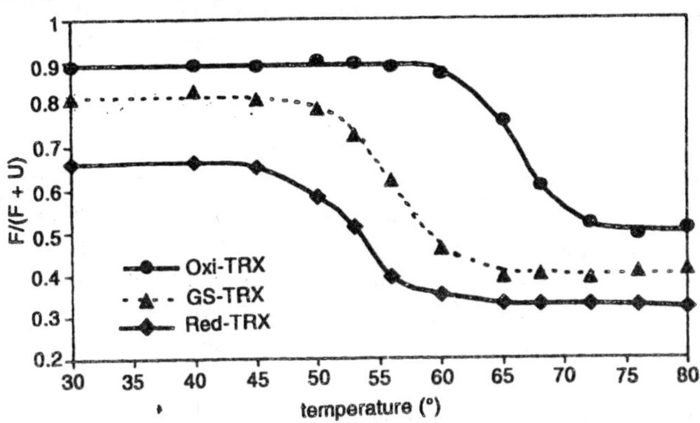


*Figure 5.5 : (A) Comparison of the heat-denaturation curves of oxidized, reduced, and (S-(2- Cys[32]ethyl)-glutathione)thioredoxin (GS-TRX) in 1 (v/v) % acetic acid deduced from the temperature-dependent charge-state distribution observed by electrospray ionisation–mass spectrometry.*

ESI-MS of allowing simultaneous assessment of conformational heterogeneity and metal ion binding. Alterations of charge-state distributions on metal ion binding/release were correlated with spectral changes observed in far- and near-UV circular dichroism (CD) and intrinsic tryptophan fluorescence. A drastic shift to higher charges was observed for the charge - state envelopes upon collapse of the tertiary structure of $Zn^{2+}$- bound colicin E9 endonuclease by acid -induced release of the metal ion. Moreover, these studies revealed that the nonc ovalent protein–protein complexes, that is, coli cin E9 endon uclease and its cognate immunity protein, Im9, dissociate in the gas phase before the metal ion complex, thereby confirming that electrostatics and not hydrophobic interactions are more important in the gas phase.

Another example that charge-state analysis can contribute to elucidate structural properties of a protein, where other methods fail, is illustrated by studies on human recombinant macrophage colony-stimulating factor-β (rhM-CSFβ). rhM-CSFβ is a four-helical bundle cytokine and forms a compact homodimer with nine disulfide bonds. The ESI mass spectrum of rhM-CSFβ exhibits a charge-state distribution centered around the 21+ charge state. Selective reduction of two disulfide bonds ($C_{157,\ 159}$–$C'_{157,\ 159}$) near the C-terminal tail showed little change in the secondary structure as suggested by far-UV CD and fluorescence spectra, and even the three-dimensional integrity of the protein was unaltered as suggested by the unchanged biological activity. However, the ESI spectrum of this partially reduced protein showed a charge envelope centered around the 32+ charge state, suggesting additional basic sites became accessible upon removal of the two disulfide bonds. However, only $Lys^{154}$ would appear to become available for protonation upon reduction of the disulfides, $C_{157,\ 159}$–$C'_{157,\ 159}$, as all other basic sites are in the N-terminal direction relative to the next stabilising disulfide bond, $C_{102}$–$C_{146}$. Constructing a model for rhM-CSFβ helped rationalising the observed charge-state alteration upon reduction. In the model, the crystal structure of rhM-CSFα was extended to aa 160, thereby providing the symmetrical $C_{157}$–$C_{159}$ disulfide linkages in the dimer. The resulting structure was subjected to molecular dynamics and energy optimisation by AMBER force field, and the remaining sequence to $Glu^{177}$ was then modeled as a turn and a polyproline type II helix. H/D exchange experiments combined with computational support suggested a structure model in which disulfide bond reduction exposes additional basic sites in each subunit, which are buried by the polyproline type II helix in the intact rhM-CSFβ.

## Negative Ion ESI-MS Charge-State Distributions

Only a few studies have been published in which the negative ion mode ESI-MS was employed to study conformational changes in proteins. In the negative ion mode ESI-MS, proteins are measured as multiply negatively charged protein species, that is, $(M - nH)^{n-}$. Buffer systems that are commonly employed to analyse proteins in the negative ion mode include ammonium acetate/ammonia or ammonium bicarbonate solutions.

An example of the use of negative ion ESI-MS involved a study in which the influence of point mutations on the global stability of wild-type ferredoxin from the cyanobacterium Anabaena was elucidated. Alterations within the charge-state distributions of the multiply deprotonated molecular ions were rationalized by a loss of structural stability due to disruption of stabilising hydrogen bonds by some of the amino acid replacements.

Another example in which negative ion ESI-MS proved useful was the report on $Ca^{2+}$-binding in calbindin $D_{28K}$ and two deletion mutants lacking one or both EF-hand domains. They compared changes in the charge-state distribution of multiply deprotonated molecular ions observed in the negative ion mode ESI mass spectra with changes in the near- and far-UV CD and the extrinsic fluorescence of calbindin $^{D}28K$ and mutants upon uptake of $Ca^{2+}$ ions. Good correlation was found between conformational changes detected by spectroscopic methods, such as near- and far-UV CD and extrinsic fluorescence, as well as changes within the charge-state distributions of the multiply deprotonated molecular ions.

However, a study by Konermann and Douglas revealed that protein conformational changes detected by spectroscopic methods and positive ion mode ESI-MS did not necessarily result in changes of the charge-state distribution in the negative ion mode. For example, acid-induced unfolding of cytochrome c and ubiquitin, as well as the base-induced unfolding of ubiquitin, led to dramatic changes in the charge distribution observed in the positive ion mode ESI-MS, whereas only minor changes were observed in the negative ion charge-state distribution. To fully judge the potential of negative ion mode ESI-MS as a method for monitoring conformational changes of proteins in solution, we must acquire more experimental data. A better understanding of the underlying processes of negative ion ESI will also be necessary.

### *Protein Folding Kinetics by "Time-Resolved"-ESI-MS*

Konermann and Douglas introduced "time-resolved" ESI- MS to study protein folding=unfolding reactions by ESI-MS. A continuous flow mixing apparatus is directly coupled to an ESI mass spectrometer. This experimental setup was first used to monitor the refolding of acid-den atured cytochrome c. The analysis of the different charge-state distributions recorded during the time course of the folding experiment revealed that refolding of acid-denatured cytochrome c proceeds via a fast folding population and a slow folding subpopulation with lifetimes of 0.17 and 8.1 s, respectively. In the second study, time-resolved ESI-MS was applied to study the acid-induced denaturation of myoglobin. The different charge-state distributions observed by ESI-MS in the time course of the unfolding experiment revealed the previously unknown transient formation of a folding intermediate with a lifetime of the order of 0.4 s, namely an unfolded state that was still capable of retaining the heme.

Using an online rapid mixing in conjunction with ESI-TOF MS, Wilson et al. have also published a study on the subunit disassembly and unfolding of the inducible nitric oxide synthase oxygenase domain ($iNOS_{COD}$) following a pH jump from 7.5 to 2.8. During the denaturation process, various protein species become populated, which were distinguished by their ligand-binding behaviour and the different charge-states distributions. The denaturation process can be described as follows: The first step is the disruption of the $iNOS_{COD}$ dimer, to generate heme-bound monomeric species in various degrees of unfolding. This first step is accompanied by the loss of two tetrahydrobiopterin cofactors. Subsequent heme loss generates

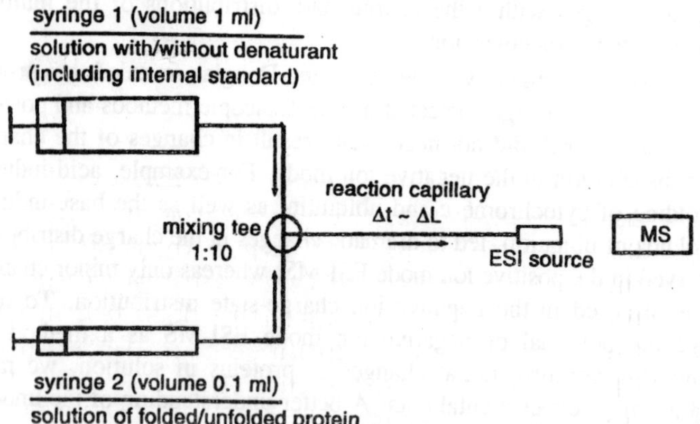


*Figure 5.6 : Experimental apparatus used for monitoring protein folding=unfolding studies by time-resolved electrospray ionisation (ESI)-mass spectrometry (MS). Two syringes are advanced simultaneously by a syringe pump. The mixing ratio is defined by the volume ratio of the syringes.*

monomeric apoproteins exhibiting various degrees of unfolding. The formation of proteins that are bound to two heme groups was also observed. Apparently, a subpopulation of holo monomers undergoes substantial unfolding while retaining contact with the heme cofactor.

## PROBING PROTEIN SOLUTION STRUCTURES

### Hydrogen Exchange Chemistry

The chemistry of hydrogen exchange reactions has been described by Englander and Kallenbach. In peptides and proteins, the exchangeable hydrogens are the polar side-chain hydrogens bound to heteroatoms (N, O, and S), the N- and C-terminal hydrogens, and the backbone peptide amide hydrogens. In proteins, the polar side-chain hydrogens exchange $10^3$–$10^6$ times faster than peptide amide hydrogens at pH 7 and exchange rate constants are highly pH dependent. Exchange rates for polar side chain hydrogens, as well as for N- and C-terminal hydrogens are usually too high to be readily determined. Therefore, most of the hydrogen exchange studies focus on determining the exchange characteristics of peptide amide hydrogens. Fortunately, peptide amide hydrogens represent excellent structural probes that report on surface accessibility, secondary structural bonding network, and conformational stability.

#### *The Influence of Solution Temperature, pH, and Composition on Amide Hydrogen Exchange*

Peptide amide hydrogen exchange is almost exclusively catalysed by OH down to pH 3 and below that by $H_3O^+$ in water-based solutions. Thus, the exchange rate constant for a freely exposed peptide amide hydrogen $k_{ex,NH}$ can be approximated as

$$k_{ex,NH} = k_{OH}[OH^-] + k_H[H^+] + k_0, \qquad (2)$$

where $k_{OH}$, $k_H$, and $k_0$ are the rate constants for base-catalysed, acid-catalysed exchange, and direct exchange with water, respectively. Amide H=D exchange studies using poly-DL-alanine as a random coil-like model peptide provided reference rate constants for $k_{OH}$ = $1.12 \times 10^{10}$ and kH = 41.7 $M^{-1}$ $min^{-1}$, respectively at 20° and low salt concentrations. Direct exchange with water is commonly considered as insignificant in most of the exchange studies ($k_0$ = 0.03 $min^{-1}$).

There is a pH dependence of exchange rate constants of polar side-chain hydrogens and exposed peptide amide hydrogens. The $\log(k_{ex,NH})$ versus pH plot has a pronounced minimum at pH 2.5–3.0, indicating that exchange rates for base and acid catalysis are equal.

The extreme pH sensitivity of peptide amide hydrogen exchange is apparent by an approximate 10-fold increase of the exchange rate for each unit change in pH.

Hydrogen exchange rates are temperature dependent. The rate constants follow the Arrhenius equation, lnk = lnA – $E_a$=(RT). Thus, to predict exchange rate constants at different temperatures the following equation can be used:

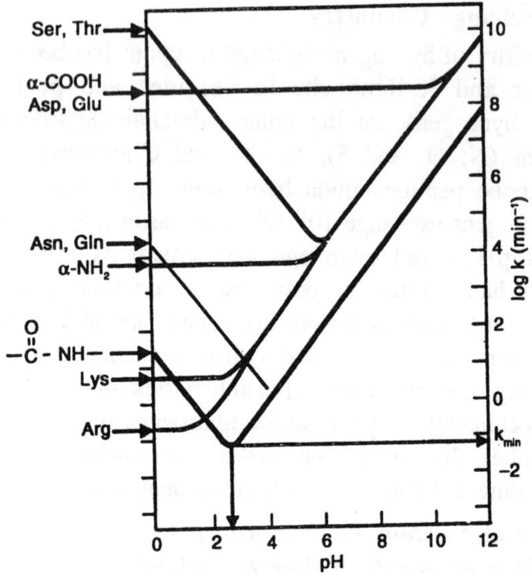


*Figure 5.7 : The pH dependence of exchange-rate constants of different types of hydrogens occurring in peptides and proteins.*

$$k_{ex,T2} = k_{ex,T1} \exp(-E_a\{1/T_2 - 1/T_1\}/R) \qquad ð3Þ$$

whereby $k_{ex,T2}$ and $k_{ex,T1}$ are the exchange rate constants at temperatures $T_2$ and $T_1$ (in Kelvin), R is the gas constant 8.314 $Jmoi^{-1}K^{-1}$ and the activation energies Ea for the acid catalysed, the base-catalysed, and water-catalysed exchange are 14 kcal/mol, 17 kcal/mol, 19 kcal/mol, respectively (1 kcal = 4.184 kJ). This corresponds to an approximate threefold increase of the exchange rate with every 10-K increase in temperature.

Temperature and pH adjustments are used in exchange rate studies of peptide amide hydrogens to (a) probe fluctuations of conformational states, (b) monitor conformational alterations as a consequence of ligand binding or environmental changes, and (c) monitor the cooperativity of unfolding transitions. At pH 7 and 25°, half-lives of exposed peptide amide hydrogen are in the range of 0.05–0.01 s, whereas at

0° and pH 2.7 the half-lives are 1–2 hr. The latter represent conditions used for quenching exchange reactions to preserve the isotope labels during analysis.

In addition to the effects of temperature and pH, solvent composition has a decisive effect on amide hydrogen isotope exchange rates. Miscible organic solvents used in reverse-phase HPLC (RP-HPLC) generally slow down the exchange rate, largely because of a decrease in the equilibrium constant $K_w$ for water or hydroxide ion activity. The $pH_{min}$ shifts to progressively higher values with increasing organic solvent fraction. A change in solvent composition can be used to reduce back-exchange of the isotopic label when working outside the $pH_{min}$ limits, for example, during HPLC analysis.

**Side-Chain Effects on Amide Hydrogen Exchange**

Peptide amide hydrogen exchange rates are influenced by neighboring side chains through inductive and steric blocking effects. Polar side chains withdraw electrons, thereby increasing the acidity of the peptide hydrogen group. Thus, the exchange rate by base catalysis is increased and the exchange rate by acid catalysis decreased. Steric blocking, and hence, retardation of exchange, is caused by bulky side chains, such as β- and γ-branching of side chains and by aromatic residues. The additivity of the inductive and steric blocking effects on peptide hydrogen exchange has been demonstrated and quantified by Bai et al. and Molday et al. through amide H/D exchange kinetics in dipeptide models of all 20 naturally occurring amino acids. These values are commonly used to calculate the exchange rate constant $k_{rc}$ for a peptide amide hydrogen imbedded in a particular amino acid sequence assuming random coil-like behaviour. The ratio $k_{rc}/k_{obs} = P$, where P is the protection factor and $k_{obs}$ is the experimentally determined rate constant for the peptide amide hydrogen in question. The protection factor enables an assessment of the retardation factor of the amide hydrogen exchange caused by secondary and tertiary structural hydrogen bonding and/or solvent inaccessibility.

**Hydrogen Exchange Mechanism: Ex1 and Ex2**

Since its introduction in the early 1950s, hydrogen exchange has been a powerful method for studying protein structure and conformational stability. Labile hydrogens in the side chains (e.g., O-H, S-H, and N-H), N- and C-terminal hydrogens, and the backbone amide hydrogens of proteins are exchangeable. However, information regarding structure and conformational stability can best be extracted from the exchange behaviour of the backbone amide hydrogens, because these build up the secondary structural bonding network. The rate at which a peptide amide hydrogen exchanges with solvent provides information about the structural stability and surface accessibility of that specific peptide amide hydrogen. Protection against exchange results predominantly from hydrogen bonding, but solvent exclusion of amide hydrogens buried in the core of a protein contributes as well. Labile protons on the surface of the protein that are not involved in hydrogen bonding exchange more readily. A dual pathway model for hydrogen exchange of peptide amide hydrogens is most often considered. The first pathway describes exchange directly from the folded state via local transient opening reactions or "breathing" motions. It is assumed that peptide amide hydrogens located at or near the surface of the protein or in close proximity to solvent channels exchange via this pathway. In the second mechanism, exchange is preceded by transient reversible unfolding reactions that are accompanied by the breaking of hydrogen bonds. Unfolding of proteins can occur via (a) global or whole molecule transitions, (b) subglobal transitions, or (c) local fluctuations. Which of the two exchange mechanisms dominates depends on the experimental conditions (i.e., the pH, temperature, or chemical denaturants present).

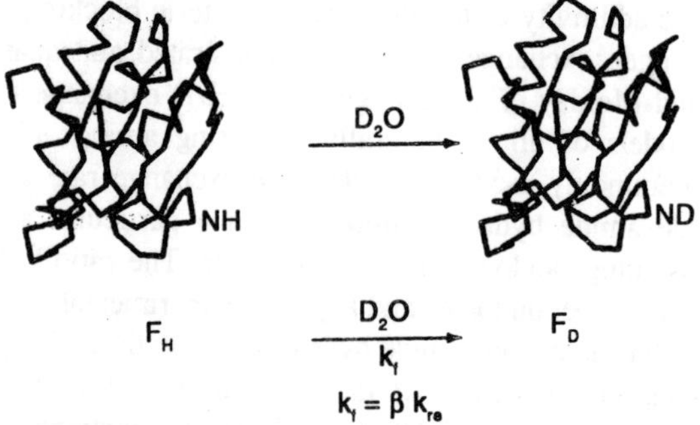


*Figure 5.8 : Pictorial presentation of the underlying principle of peptide amide hydrogen exchange from folded proteins. The rate constant for exchange at any backbone amide is given by $k_{obs} = \beta k_{rc,}$ where $\beta$ is the probability that $D_2O$ and $OD^-$ are present at the back amide and $k_{rc}$ is the intrinsic chemical rate constant for isotopic exchange at amide hydrogens in random coil like peptides.*

The Linderstrøm-Lang model connects the second mechanism with the observed exchange rates.

The observed exchange reaction will be second order, if the refolding rate of the transient opening reaction is much greater than the rate with which the intrinsic chemical exchange occurs (i.e.,

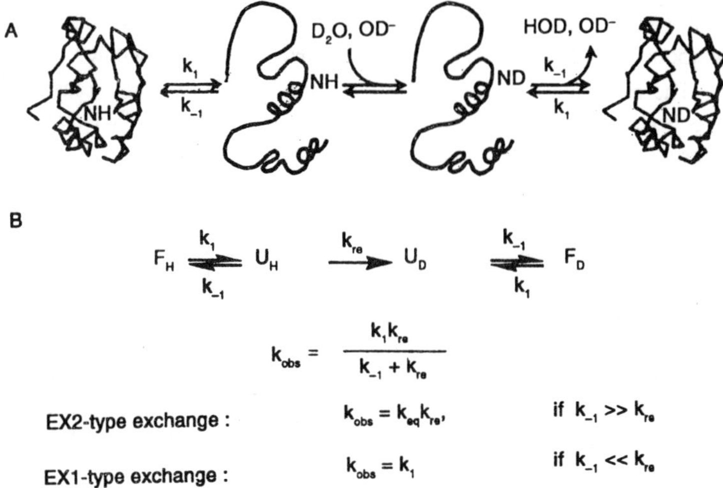


*Figure 5.9 : (A) Pictorial presentation of the underlying principles of peptide amide hydrogen exchange from folded proteins via partial, subglobal, or global unfolding. (B) The Linderstrøm-Lang model (Eq. [5]), which connects observed exchange rates with structural unfolding reactions in proteins. Rate constants $k_1$ and $k_{-1}$ describe the unfolding and refolding reaction of a protein, and $k_{rc}$ is the intrinsic chemical rate constant for isotopic exchange at amide hydrogens in random coil-like peptides.*

$k_{-1} >> k_{rc}$). Thus, the experimentally observed rate constant is given by $k_{obs} = k_{rc}$ ($k_1 = k_{-1}$). This type of exchange is termed the *EX2-type mechanism*, and exchange via this mechanism is observed for most proteins under neutral pH and in the absence of denaturants. In this mechanism, refolding (i.e., large $k_{-1}$) is more likely than the exchange reaction and the exchange is random or uncorrelated. Exchange occurs via the so-called EX1 -type mechanism, if the rate of the chemical reaction is much greater than refolding (i.e., $k_{rc} >> k_{-1}$). Thus, the observed exchange rate is directly related to the protein unfolding rate (i.e., the observed exchange rate constant is given by $k_{obs} = k_1$), and correlated motion can be detected by MS.

## Hydrogen Isotope Exchange Monitored by Mass Spectrometry

Typically, H=D exchange experiments are monitored by NMR techniques because deuterons in proton NMR experiments do not give a signal. The exchange behaviour of individual amide protons as a function of exchange periods is deduced from the intensity of resolved NH-resonances in one-dimensional (1D) NMR spectra or NH-related cross peaks in two-dimensional (2D) NMR spectra. Thus, deuterium exchange experiments in which the average proton occupancies is monitored in a site-specific manner yield exchange rates of individual backbone amides in proteins but averaged over all molecules.

The advantages of using MS to monitor H=D exchange reactions have become recognized. MS-based approaches use the fact that the difference in mass between hydrogen and deuterium is 1 Da, and, thus, changes in the number of deuteriums incorporated in a molecule can be directly related to mass shifts. The application of MS to monitor H=D exchange in peptides and proteins was first introduced by Katta and Chait to study the native and denatured states of ubiquitin. Since then, the approach has been widely used to (a) study the high-order proteins structures in different environments, (b) assess the consequences of point mutations and ligand binding, (c) probe the solution dynamics of proteins and peptides, (d) obtain structural descriptions for partially folded states, and (e) examine protein folding and unfolding pathways.

In principal, two experimental approaches are used. In the first approach, the deuterium exchange-in reaction is followed by dilution of the protonated protein in deuterating solvents. This approach is beneficial for proteins that are difficult to prepare in a fully deuterated form, but it suffers from the fact that back-exchange during the ESI-MS analysis may occur. Therefore, this methodology is more suitable for experiments in which comparisons between different proteins or conformations are made. In this context, it is important to note that if deuterium exchange-in experiments are directly monitored by ESI-MS, multiply deuterated ions $(M + nD)^{n+}$ will be observed. Thus, deuterium incorporation can be monitored directly as mass increase and expressed as the percentage of mass increase relative to the maximum number of exchangeable hydrogens.

In the second approach, the exchange is initiated from a fully deuterated protein by dilution in protonating solvent. Under these conditions, deuterium exchange-out kinetics are monitored. This process has the obvious advantage that the exchange rates observed are directly related to the exchange behaviour of the backbone amide deuterons because exchange of the side-change deuterons is several orders of magnitude faster than exchange from the backbone amides. Fortunately, the exchange behaviour of the backbone amides is of particular interest for structural studies, because the backbone amides provide the framework for the hydrogen bonding network that build up the tertiary structure. Under these circumstances, if deuterium exchange-out experiments are directly monitored by ESI-MS, multiply protonated $(M + nH)^{n+}$ ions are observed and the mass decrease as exchange-out proceeds can be best expressed as the number of deuteriums remaining after a certain exchange-out period. The deuteriums remaining can be determined by subtracting the observed mass shift (i.e., the deuterium loss) from the number of deuteriums observed in the fully deuterated protein.

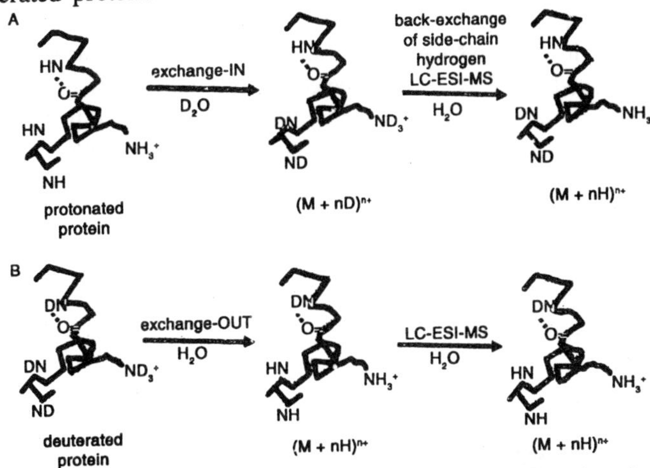


*Figure 5.10 : The two experimental approaches used to perform isotopic exchange studies: (A) the exchange-in and (B) the exchange-out approach.*

Monitoring and analyzing D=H exchange-out is particularly powerful for studying protein interfaces. The protein–protein interfaces are indicated by regions that retain more deuterons in the complex compared with control experiments in which only the individual, non-complexed protein is present. Monitoring exchange-out was used, for example, by Mandell et al., for the identification of the kinase inhibitor PKI (5–24) and ATP-binding sites in the cyclic AMP–dependent protein kinase. The same authors have also published the identification of the thrombin–thrombomodulin interfaces by monitoring the kinetics of amide deuterium exchange-out.

The influence of a disulfide bond on the thermodynamic stability of a small single-domain protein can be seen in deuterium exchange-in experiments, with oxidized and reduced Escherichia coli thioredoxin (TRX) using ESI-MS to measure the mass increase. Oxidized *E. coli* TRX possesses 173 exchangeable hydrogens; 102 are backbone amide hydrogens, 68 are side-chain hydrogens, and 3 are terminal hydrogens. The reduced protein has a 175 exchangeable sites because of two additional thiol protons. For oxidized TRX, a molecular mass increase of approximately 71 ± 5 Da (40 ± 3%) is observed after 60-min incubation in 1% acetic acid-d1 (v=v) at room temperature, indicating that approximately 102 ± 5 of the hydrogens did not yet exchange, and these are protected against exchange and, presumably, are part of the stable protein core. By contrast, in reduced TRX, approximately 82 ± 5 (47 ± 3%) protons have undergone exchange, leaving approximately 93 ± 5 hydrogens on the protein. A comparison of the global hydrogen exchange profiles over a 2-h period demonstrated that reduced TRX is less protected against deuterium exchange-in over the entire time course sampled, thereby confirming that reduced TRX is thermodynamically less stable than oxidized TRX.

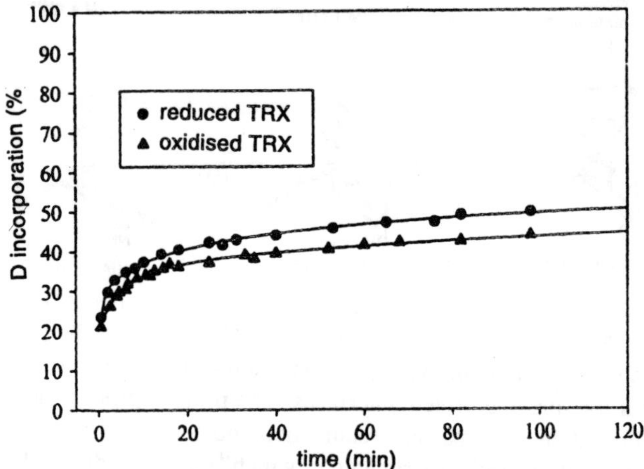


*Figure 5.11 : H/D exchange-in kinetic profiles of reduced and oxidized Escherichia coli thioredoxin (TRX) at 22° in 1% acetic acid-d1 (v/v).*

Zhu et al. introduced a methodology to quantify protein-ligand interaction by MS, ligand titration, and H=D exchange, termed PLIMSTEX. This approach can determine conformational change, binding stoichiometry, and affinity in protein–ligand interactions including those that involve small molecules, metal ions, and peptides. The method yields plots of the mass difference between the deuterated and nondeuterated protein (deuterium uptake) versus the total ligand concentration. These titration curves are then fit to a 1:n (n is the number of binding sites) protein:ligand sequential binding model. This method was applied to determine the $Ca^2$ interactions in calmodulin and to study the interaction of the intestinal fatty acid-binding protein (IFABP) with a fatty acid carboxylate.

## Determination of H/D-Exchange Mechanisms: EX1 and EX2 Types

Proteins are not rigid structures but highly dynamic systems that undergo constant fluctuations in which hydrogen bonds are transiently opened and re-formed. The extent of these fluctuations is critically dependent on the environment that can stabilize or destabilize the native conformational state. Thus, backbone H/D exchange in proteins under solution-phase conditions depends on two processes: (a) the unfolding or opening event during which the backbone amides are transiently exposed to the solvent and (b) the intrinsic chemical exchange of the exposed amide hydrogen with deuterating solvent. The two rate-limiting mechanisms (i.e., EX2 and EX1 types for isotopic exchange reactions in proteins) can be distinguished. When a large excess of the hydrogen isotopically labeled water is present, the overall exchange reaction is essentially irreversible.

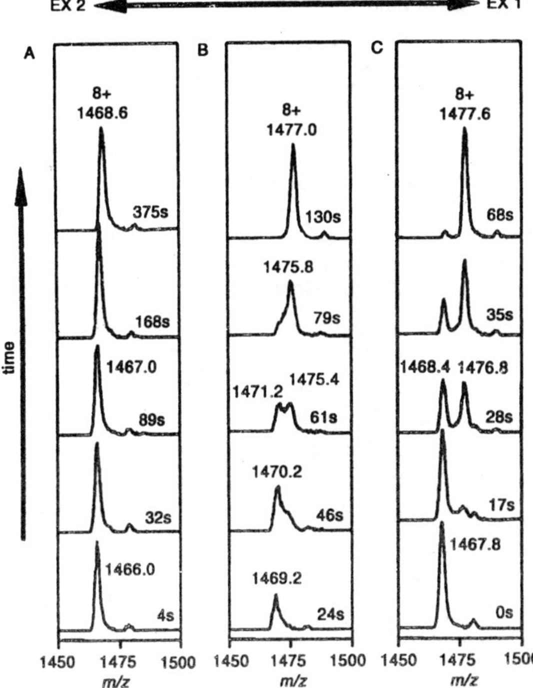


*Figure 5.12 : Evolution of the eightfold charged ion peak of Escherichia coli thioredoxin during the online H=D exchange-in experiments at 40° (A), 60 ± 2° (B), and 80° (C) in 1% acetic acid-d1 (v=v). The time points given refer to incubation periods (time).*

The conformational dynamics of TRX during thermal unfolding have been studied by solution-phase H/D exchange-in as a function of time and temperature. To study the dynamic processes of TRX by H/D exchange-in during heat denaturation, mild acidic solution conditions were chosen to slightly destabilize the native state. H/D exchange was initiated by diluting the protein solution into a deuterating solution, which consisted of 2% acetic-acid-$d_1$ (v/v). A capillary assembled in a conventional injection valve was used as a reaction vessel. The solvent delivery line, the reaction vessel, and the injection valve were immersed in a water bath equilibrated at the desired temperature. Thermal denaturation and H=D exchange-in reactions were performed in the reaction vessel that was directly connected by a fused silica capillary to the ESI source. This infusion capillary was immersed in an ice bath to quench H=D exchange-in and initiate refolding of thermally denatured TRX.

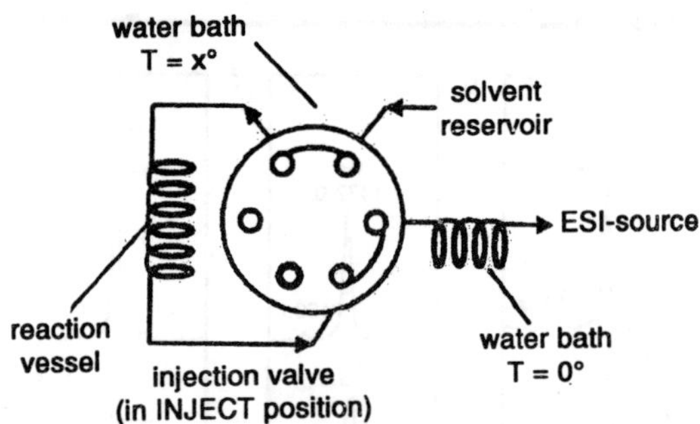


*Figure 5.13 : Experimental setup used for the H=D exchange in experiments monitored online by electrospray ionisation–mass spectrometry.*

Exchange according to a pseudo–first-order mechanism (EX1-type mechanism) occurs if interconversion between the native closed conformational states (F) and the unfolded open states is slow with respect to the intrinsic hydrogen exchange rate. In this case, the rate limiting step is determined by $k_{-1}$ and the experimentally observed exchange rate constant $k_{ex}$ is directly related to the unfolding rate constant k1 (i.e., $k_{ex} = k_1$). These are normal denaturing conditions (e.g., high temperature or in the presence of chaotropic reagents) under which the native state is only marginally stable. Because the chemical exchange rate is highly pH dependent, alkaline pH favours exchange according to an EX1 type of mechanism. Mass spectrometric approaches are uniquely able to detect the cooperativity of exchange of multiple sites because the correlated exchange of several sites results in a mass difference, which results in a bimodal ion peak pattern. Such a scenario is observed during thermal denaturation coupled with deuterium exchange in experiments of oxidized TRX above 65°. After a short incubation period, a bimodal ion peak pattern was observed for each charge state detected. The ion peak at lower m=z values represents the population of protein molecules that had not undergone cooperative unfolding, and their core amide protons were still protected against exchange in, whereas the ion peak at higher m=z values arises from the deuterated population of molecules that had undergone cooperative unfolding. Hence, these molecules had spent enough time in the unfolded state to allow their core amide protons to exchange. The number of amides that participate in the cooperative unfolding event can be extracted from the m/z difference of these two ion peaks under consideration of the charge state. In the case of oxidized TRX, 68 ± 6 deuterium underwent correlated exchange. If exchange occurs exclusively according to an EX1 mechanism, the intensity of the ion peak at lower m/z values decreases during exchange-in, whereas the intensity of the ion peak at higher m/z values increases, in which case no mass shift is observed. The rate of conversion allows an estimation of the unfolding rate constant $k_1$. For oxidized TRX in 2% acetic acid-d1 (v/v) at 80°, approximately 68 ± 6 backbone amides underwent cooperative unfolding with a rate constant k1 of approximately 2 ± 0.2 min.

Under conditions that favour the compact native state, exchange occurs via a second-order mechanism (EX2-type mechanism) that is characterized by a fast interconversion between native and unfolded open states. Thus, the overall exchange rate of the backbone amide hydrogen depends on the intrinsic exchange rate and the average fraction of time during which the backbone amide hydrogen is transiently in the open state and exposed to bulk solvent (i.e., $k_{obs} = k_{rc} (k_1/k_{-1})$. H/Dexchange behaviour of the EX2 type has been observed for oxidized TRX when exchange experiments are performed below 40°. Under these conditions, the ESI mass spectra showed ion peaks that shifted gradually to higher m/z values with increasing exchange-in time because each amide will exchange to a varying degree depending on its individual exchange rate.

Another example that demonstrates the beauty of using MS-based H/D-exchange approaches to investigate conformational dynamics is the study on E. coli heat shock transcription factor $\sigma^{32}$ by Rist et al. Using an HPLC-MS setup, they monitored the incorporation of deuterons into full-length $\sigma^{32}$. In combination with immobilized pepsin for digestion, these authors localized slow- and fast-exchanging regions within the entire sequence of $\sigma^{32}$. These studies revealed that $\sigma^{32}$ adopts a highly flexible structure at 37°, as indicated by a rapid exchange of about 220 of the 294 amide hydrogens. However, at 42°, a slow-correlated exchange of 30 additional amide hydrogens was observed. The domain that exhibited correlated exchange was part of a helix-loop-helix motif within domain sigma 2, which is responsible for the recognition of the —10 region in heat shock promoters. The correlated exchange is related to reversible unfolding with a half-life of about 30 min due to a temperature-dependent decrease in stabilisation energy. These data support a role for $\sigma^{32}$ as a thermosensor by undergoing conformational alterations in response to heat shock.

Finally, it is important to point out that mass spectrometric approaches are uniquely able to distinguish between EX1 and EX2 exchange regimens because the distribution of deuteriums can be obtained from the mass spectrum. NMR-based approaches are not able to directly distinguish between these two exchange mechanisms because NMR methods monitor the average proton occupancy at individual sites. A number of authors have emphasized the complementary use of both techniques (i.e., NMR and MS) to investigate the native-state H=D-exchange experiments of small single-domain proteins. In the latter study, the third domain of turkey ovomucoid was subjected to native-state H=D-exchange experiments. NMR studies suggested that 9 of 14 of the slowest exchanging peptide hydrogens possess very similar unfolding rate constants and free energies, suggesting that the exchange of these 9 backbone amide hydrogens participates in a cooperative unfolding event. However, ESI-MS data and mass peak simulations revealed that only three to five (not nine) backbone amide hydrogens exchange cooperatively. In this case, the combined use of NMR and MS to study native-state amide hydrogen exchange resulted in a very detailed picture of the thermodynamic and kinetic properties of the very slowest exchanging backbone amide hydrogens of the turkey ovomucoid domain. This kind of knowledge is important for understanding the nature of conformational fluctuations and cooperativity of folding=unfolding events in the modern view of funnel-like energy landscapes of proteins.

## Peak Width Analysis

The peak width of a molecular ion in an ESI mass spectrum of a protein arises from the contribution of the natural isotope, primarily that of carbon-13 abundance to the individual charge state and the instrumental resolving power. If mass spectrometric measurements are done on small proteins, the ion peak width of a certain charge state at half height compares well with the natural peak width simulated. For example, the peak width at half height of approximately 1.2 amu observed for the eightfold protonated molecular ion for oxidized TRX compares reasonably well with the theoretically expected peak width at half height of approximately 1 amu for the eightfold protonated ion peak of oxidized TRX, which has the empirical formula $C_{28}H_{836}N_{132}O_{159}S_3$.

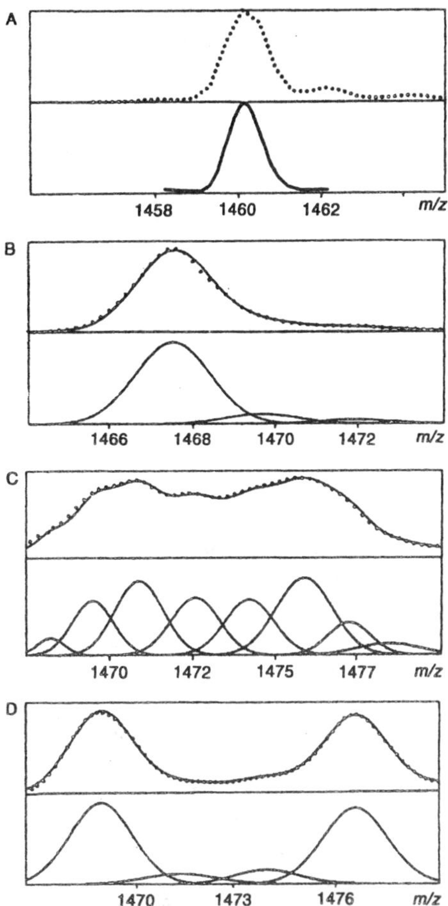


*Figure 5.14 : Peak width analyses of the eightfold charged ion peak of Escherichia coli thioredoxin observed during the exchange in experiments under different exchange conditions.*

In experiments in which MS is used to monitor H=D exchange, the analysis of the ion peak width may reveal unique information regarding the conformational diversity of the protein and the exchange kinetics. A consequence of H=D-exchange experiments is a broadening of the ion peak caused by the convolution of the natural isotope distribution with an assumed random distribution of deuterium throughout the protein. For example, during H=D experiments under EX2 conditions, the peak width at half height of the eightfold protonated molecular ion of oxidized *E. coli* TRX was approximately 2.5 amu throughout the time course of the exchange-in experiments, which suggests that a relatively small distribution of conformational states is present and that peak broadening is primarily caused by random, uncorrelated deuterium incorporation via local unfolding events.

In contrast, when the exchange experiment was performed at elevated temperature, 60°, the appearance of the ion peak changed dramatically after only short incubation periods. There was significant broadening of the observed ion peaks. For instance, for the eightfold protonated ion peak of oxidized *E. coli*, TRX showed a width at half height of approximately 8 amu. This type of broadening signifies that under conditions favouring the transition zone between the EX1 and EX2 mechanism, a wide range of conformational states with differing extents of protection against exchange-in exist.

As exchange-in experiments are performed under conditions favouring the EX1-type mechanism, two distinct ion peaks evolve for each charge state, indicating that the compact folded and unfolded states are simultaneously populated. For example, the peak widths at half height for the eightfold protonated ion peaks representing the folded and unfolded population are approximately 2.5 amu, which can be interpreted on the basis that the folded and the unfolded conformational states are represented as distinct populations of molecules. Gaussian deconvolution of the ion pattern reveals two minor additional populations (<10 %) exhibiting deuterium incorporation that might be expected for partially unfolded states. In addition, during the course of the exchange-in at elevated temperature, a slight shift to higher m=z values and a concomitant narrowing of the ion peak at higher m=z value were observed, which indicates that the exchange mechanism was not exclusively EX1, but a combination involving both the EX1 and the EX2 type of mechanism. Yi and Baker and Arrington et al. have described an algorithm that used NMR-derived exchange rate constants to simulate accurate binomial mass spectral data under consideration of a different contribution of the two rate-limiting exchange mechanisms (i.e., EX1 and EX2), various degrees of cooperativity, and the natural isotope distribution.

**Regional Structural and Dynamic Information from Medium Spatial Resolution Data**

In addition to probing the global conformational stability and dynamics of a protein, more detailed information with a medium spatial resolution can be obtained by digesting the deuterium-labeled protein with pepsin under conditions of slow exchange (i.e., pH 2.5 and 0°) and then analyzing the resulting peptic fragments by liquid chromatography online- coupled MS. Both fast-atom bombardment (FAB) and ESI-MS have been used. The deuterium levels of the individual peptides can be extracted from the observed mass shifts between the unlabeled and labeled peptides. This approach has been used in numerous applications to characterize the higher order structure and dynamics of proteins or to study the protein conformational influence on ligand binding. The protein proteolysis approach is especially attractive because of its universality, but it is relatively time consuming and suffers from increased probability of back-exchange of hydrogen isotopes during chromatographic separation and the limited proteolysis often observed for protease-resistant proteins. Procedures to adjust for deuterium loss during the analysis have been published. Other authors have described HPLC setups that combine online pepsin digestion and LC-ESI-MS analysis.

An example in which conformational and solvent accessibility changes were studied by H=D exchange, peptic proteolysis, and MS is the study of human retinoid X receptor homodimer upon 9-cis retinoic acid binding. The RXRci LBD is a 54.4-kDa non-covalently linked homodimeric protein. Time course labeling studies in the presence and absence of its ligand 9-cis retinoic acid were performed. Deuterium exchange levels with medium spatial resolution were obtained by peptic proteolysis followed by mass analysis. The deuterium exchange profiles of peptic peptides generated from the apo- and holo-RXRci LBD homodimer were compared. It became apparent that approximately half of the regions that show significant exchange protection are regions that have direct contact surfaces with the ligand, as indicated by the X-ray structure: The amino termini of helixes 3 and 9, the two β-sheets, helix 8, the H8-H9 loop, and the C-terminus of helix 11. Unexpectedly, protection was also observed in peptides derived from helixes 7, 10, 11, and the H7-H8 and H10-H11 loops, regions that are not directly in contact with bound 9-cis-RA, and these changes were not apparent in the X-ray structure of the holo protein. Similarly, the ability of MS-based H/D exchange, in combination with proteolysis to determine both the location of binding sites and the effects of binding on distal locations, was also reported for different states of actin.

**Continuous Vs. Pulse Labeling**

Essentially, two experimental H/D-exchange approaches are used: continuous exchange and pulse labeling. Continuous exchange involves exposing the protein to the deuterated medium during the incubation time. The labeled protein is thus a measure of all unfolding events that took place, provided the exchange rate is much faster than the unfolding event. In pulse labeling, the deuterated protein is a snapshot of the unfolded protein that existed during the pulse. Many unfolding and refolding events may have taken place during the incubation, but these are not recorded because no deuterium is available for exchange between the pulses.

Proteins can undergo H/D exchange either from their folded states in which case the exchange involves amide hydrogens at the surface of the protein that are not hydrogen bonded, or they can exchange after unfolding, which means that hydrogen bonds were broken in order for the exchange to take place. A number of authors have shown that the combi nation of continuous versus pulse labeling in $D_2O$ can help to reveal when both mechanisms are operative in different regions of the protein. The approach is particularly useful when labeling under denaturing conditions occurs to give bimodal isotope distributions. Thus, for rabbit muscle aldolase in 3 M urea, when exchange in a given region of the folded state was slow, a bimodal isotope distribution displayed in the mass spectra of peptic peptides derived from the intact protein was observed in both experiments, but the relative intensity of the protonated distribution by pulse labeling diminished more slowly with time. When exchange was rapid, poorly resolv ed isotope distri butions observed under continuous exchange conditions in the examined peptides of certain regions of the protein were nicely resolved into bimodal distributions in the pulsed labeling experiment.

**In Source Collisionally Activated Dissociation of Proteins**

There are two basic ESI dissociation strategies one takes place in the ESI source, between the end of the capillary and the skimmercone, and the other takes place after the ESI source. In the latter, ions are sele cted for CID and the products are then analysed by standard MS /MS strategies. This procedu re is commonly used for sequencing peptides. In the former, collision - activated dissociation (CAD ) of protein ions is induced in the ESI interface region by a large voltage difference between the capillary and the skimmer cone. This method is a useful alternative to the usual procedure of digesting the protein by proteol ytic enzymes—in the case of H/D exchange experiments by pepsin— and introducing the digest into the mass spectrometer by infusion or HPL C. Fragmentation by CAD eliminates the sample workup time and reduces the amount of sample required. Solution-phase H/D exchange and CAD in the nozzle - skimmer region of the electrospray source in combination with ESI-FT-ICR MS was used to probe the conformational stability of cellular retinoic acid-binding protein 1. A non structured 22 - residue N - terminal His tag underwent complete amide isotope exchange, while the peptide fragments from regions of stable β-sheets and cr- helixes were found to be well protected against exchange.

### High-Resolution Mass Spectrometry (FT-ICR-MS) for H/D Exchange Studies

The combination of electrospray ionisation with Fourier transform ion cyclotron resonance (FT-ICR) MS allows for the analysis at unit mass resolu tion of pro teins whose mass es exceed 40 kDa. By doubl e de pletion of $^{13}C$ and $^{15}N$ isot opes to less than 0.0 1%, the mass range at which unit mass resolution is observable can be extended to nearly 100 kDa. Unambiguous assignment of charge states is a key advantage to high-resolution FT-ICR-MS. Studies of gas-phase conformational states and changes between states and protein folding by the H/D exchange methodology depend on the accuracy of the charge - state assignments. When low - resolution MS is used, the determination of charge states may require HPLC separation of the peptides. Such separation schemes take time and increase the probability for back-exchange of the isotope. Identification of the peptic peptides or other proteolytic fragments is necessary for successful studies on protein conformations by H/D exchange; and the identification of these peptides can readily be achieved in most cases by accurate mass determinations alone. Low-resolution MS may require MS/MS experiments, which take time.

## TANDEM MASS SPECTROMETRY AND SITE-SPECIFIC PROTEIN AMIDE HYDROGEN EXCHANGE

### The $b_n$-Ions

In comparison to NMR methods, MS offers certain advantages for probing conformational structures. The one advantage enjoyed by

NMR was the ability to determine deuterium levels at the amide nitrogen of individual peptide linkages. Amide H/D content can also be measured by tandem MS at the individual amino acid residue level; however, the results must be interpreted with caution because of the potential for randomisation during the collision process. Deuterium present on amide nitrogens will be evident in the shift to higher masses of the isotopic clusters of sequence ions, such as the $b_n$- and $y_n$-ions, when compared to the nondeuterated peptides. From these data, the H/D ratios at individual amide nitrogens can be determined.

For these determinations to succeed, a sequential set of ions must be observed. Although earlier studies used several ion series, such as the $b_n$, $y_n$, and $z_n$, subsequent studies rely mostly on the $b_n$ -ions. A potenti al prob lem in using y n-ions is that they are believed to be formed by the transfer of a hydrogen from the N-terminal side of the cleavage site. This transfer of hydrogen or deuterium during analysis obviously would lead to errors if not accounted for. Thus, analysis of the bn-ions is preferable because no transfer of hydrogens is involved during fragmentation to the acylium ions. However, the collision process required for fragmentation in collisional-induced dissociation (CID) tandem MS (i.e., MS/MS) can cause intramolecular scrambling of the amide hydrogens. For example, Demmers et al. reported gas-phase deuterium scrambling in peptide ions of transmembrane peptides, in which the near-terminal amino acids that anchor the peptide at the lipid–water interface were systematically varied. Using nanoelectrospray and CID on a quadrupole TOF instrument, the extent of intramolecular hydrogen scrambling was influenced by the amino acid sequence of the peptide, the nature of the charge carrier, and, therefore, most likely also by the gas-phase structure of the peptide ion. Protonated $b_n$- and $y_n$-ions were evaluated, as well as sodiated a- and y-type fragment ions. Although the observed scrambling seems to be mostly independent of the peptide fragment ion type, scrambling seems to be reduced by using alkali metal cationisation instead of protonation in the ionisation process.

Randomisation occurs by the "mobile proton" (i.e., by the added proton that forms the $MH^+$). All exchangeable hydrogens are subject to interchange. Extensive H=D scrambling was observed in sustained off-resonance irradiation (SORI) CID in ESI-FT ICR - MS an alysis, although ci-helical regions were found to remain intact by salt bridge stabilisation. The degree of hydrogen scrambling depends to a large extent on experimental conditions and the nature of the peptides being analysed. High internal energy of the peptide ions in the gas phase results in decreased H=D exchange rates. Slow heating involved in SORI activation leads to extensive intramolecular hydrogen scrambling, whereas rapid heating should reduce scrambling. There is evidence that electron-capture dissociation (ECD) to fragment proteins into a series of $c_n$- and $z_n$-ions causes much less H/D scrambling, but fragmentation by ECD alone appears to yield sequence information only for the terminal regions of the peptide. To obtain a more complete ion series, or to analyse larger proteins, a combination of ECD and collisional activation by a background gas (CAD) or blackbody infrared irradiation (IR) is necessary. How this combination of ECD and CAD or IR will affect hydrogen scrambling is not yet known. Scrambling appears to be much less prevalent for short helical peptides or short peptides generally.

In addition to hydrogen scrambling during the collision process, there are some other limitations in using MS for determining the amide H=D ratios. Experience has shown that CID MS=MS will not always yield complete sequence information for peptides. Usually, intermediate-size peptide fragment ions are observed, but the smaller ones (dipeptides, tripeptides, etc.) are sometimes missing. As the size of the peptides increases, the completeness of the sequence usually diminishes, and peptides of 30 or more residues almost never yield complete sequence ions. Some information can in principle be obtained for these ions from the reverse direction sequences and possibly from other ion series, but again, the potential for hydrogen scrambling in other ions must be kept in mind. The total deuterium content in the stretch of amino acid residues representing the gaps in the sequence, which often involves no more than two or three residues, will, of course, be recorded and the average number of deuteriums per residue can be determined.

The charge state of peptide ions selected for collisional activation also is relatively important. Higher charge states of the selected peptide ions (i.e., +3 and +4) can be used , but it is often more difficult to interpret the results because of the uncertainty in the charge states in the fragmentions. In addition, errors in mass measurements are multiplied by a fact or equal to the charge. High-resolution FT- ICR instruments overcome these problems and are particularly advantageous for studying higher charged ions.

***Correlation of H/D Ratios and NMR-Derived Exchange Rates***

Despite the potential problems, recent pulse-labeling studies with cytochrome *c* suggest that it is, in fact, possible to obtain reliable

H/D exchange information on backbone amide groups of peptides by tandem MS /MS. These new results confirm the results of some limited continuous labeling studies performed earlier with cytochrome c. If it is assumed that randomisation of exchangeable amide hydrogen by the "mobile proton" is statistical, in other words no preference for any given amide site, once in the gas phase, then it is reason able that the fractional deuterium content measured at a particular amide site has some relationship to what it was before collisional activation. The studies by Deng et al. have shown that the level of deuteration of amide groups in the b-ion series and even the y-ion series following exchange -in from $D_2O$ for 10 s can be qualitatively corroborated with H/D exchange rate constants as determined by NMR. Fast exchange rates usually result in high levels of deuterium, whereas slow exchange rates result in low deuterium levels at the amide groups. These authors have concluded that intramolecular migration of deuteriums for the $b_n$-ions is negligible. The y-ions, on the other had, were found to be less reliable, as some discrepancies were observed, suggesting H/D transfer during CID fragmentation.

Studies with oxidized and reduced *E. coli* thioredoxin for which NMR exchange rate data are available show good qualitative correlations between the hydrogen rates of exchange and the amount of deuterium label at the amide sites in the $b_n$-ions, but this was not the case for the $y_n$-ions. Negative values and values greater than 1 for site-specific amide deuterium levels are likely due to hydrogen scrambling. This problem is most severe for the $y_n$-ions. To determine accurate deuterium levels on the $b_n$-ions, it is critical to obtain the centroids of the isotopic peak clusters. Algorithms have been published to calculate these values, and adjustments for deuterium loss during the analysis can also be made.

**Conformations and Site-Specific Hydrogen Bonds**

Chemical modifications of proteins cause changes that may alter the conformations, the folding and unfolding profiles, and the protein's stability toward denaturation under different pH levels and temperature conditions. A C35A *E. coli* thioredoxin mutant shows little change in the proton NMR resonances relative to the wild-type protein, although significant changes are observed in the resonances of 175 and P76, which are within van der Waals' contact of C35 in the wild type. A more significant effect from the modification of the protein is the large change in $pK_a$ of D26 due to shielding of the region from solvent by a covalently attached peptide at C32 in the mutant. Alkylation of

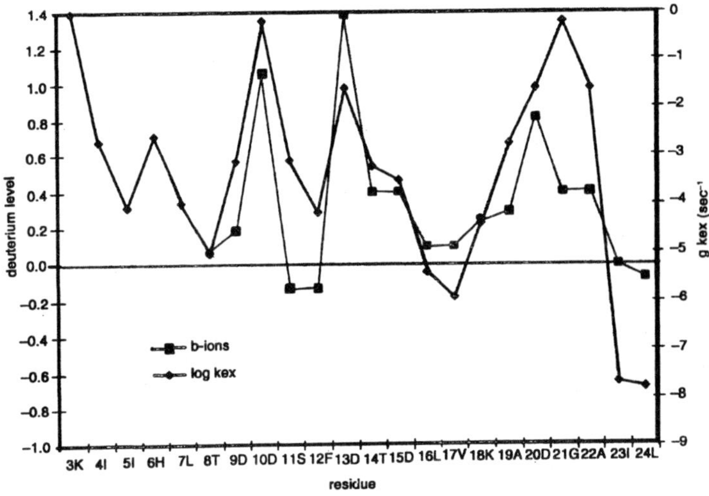


*Figure 5.15 : Comparison of the deuterium levels at amide sites in the peptide 1–24 of oxidized thioredoxin based on the analysis of $b_n$ ions and H=D exchange rate constants from nuclear magnetic resonance data.*

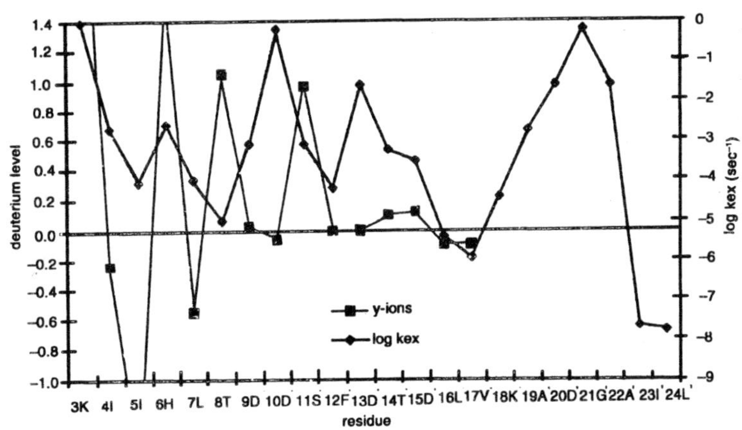


*Figure 5.16 : Comparison of the deuterium levels at amide sites in the peptide 1–24 of oxidized thioredoxin based on the analysis of $y_n$ ions and H=D exchange rate constants from nuclear magnetic resonance data.*

*E. coli* thioredoxin at Cys-32 by S-(2-chloroethyl) glutathione or S-(2-chloroethyl) cysteine yields protein structures that show a reduction in the melting temperature of the protein and an increase in the deuterium incorporated into the overall structure relative to oxidized thioredoxin. But the effects are not strictly related to the fact that a disulfide linkage is lost through alkylation, because reduced thioredoxin incorporates still more deuteriums and has even a lower melting temperature than the alkylated proteins.

The ethylglutathionyl thioredoxin structure when subjected to energy minimisation using AMBER force-field showed that the Arg-73 side chain is rotated and forms a salt bridge with the carboxylate group of the glutathione γ-Glu residue. A second salt bridge was predicted between the carboxy terminus of the glutathione Gly residue and the E-amino side chain of the Lys-90 residue. In addition, in the model the glutathione moiety forms three hydrogen bonding interactions with amide nitrogens and carbonyl oxygens of the protein backbone. The amino termini of the γ-Glu, the γ-carbonyl oxygen of γ-Glu, and the carbonyl oxygen of the Cys residue of the glutathionyl moiety form hydrogen bonds to the following moieties of the protein backbone, the carbonyl oxygen of Arg-73, the amide nitrogen of Ile-75, and the amide nitrogen of Ala-93, respectively, From these interactions, it was predicted that the amido nitrogen of Ile-75 and Ala-93 would be protected from solvent by the glutathionyl moiety and, therefore, will show a reduction in the rates of H/D exchange. The H/D exchange experiments confirmed the predictions and showed, in deed, low levels of deuterium incorporation at the backbone nitrogens of theses residues.

Deuterium content following H=D exchange on amides near the active site region of oxidized and reduced thioredoxin generally show excellent correlation with NMR hydrogen isotope exchange rate constants. In striking contrast is the deuterium content for Ile-75 and to some degree for Ala-93 for the (S-(2Cys$^{32}$)ethyl)glutathione)-thioredoxin (GSTRX), both of which show low levels of deuterium. (S-(2-(Cys$^{32}$) ethyl)-cysteinyl)-thioredoxin (Cys-TRX) does not show low incorporation of deuterium at Ile-75 and Ala-93, and there is little difference in labeling for this region of Cys-TRX protein compared to Oxi- and Red-TRX.

**Conformations in the Gas Phase**

Most H=D exchange studies for probing protein conformations have been conducted in solution. Increasing evidence indicates that proteins retain some of their conformational properties during ionisation and analysis by MS. Accordingly, there has been greater emphasis in studying proteins in the gas phase because the absence of water is expected to lead to results that correlate better with theoretical predictions. Comparisons of H/D exchange data between these two phases should provide information on the role of water in protein structures, particularly where conformational changes take place such as in the folding pathways. In the absence of water, a multitude of transient conformers result from strong intramolecular interactions that precede the final native conformation in a funnel-like fashion. FT-ICR-MS is ideally suited for probing the gas-phase conformational preferences of proteins because the ions can be trapped and then exposed to $D_2O$ for varying time periods before their excitation and detection.

To a first approximation, the number of exchangeable hydrogens of a given protein conformation is independent of charge state, despite that collision cross sections and ion mobility studies show that the diffuseness of proteins is highly dependent on the charge state. Thus, if the number of exchangeable protons changes significantly with the charge state, different conformations are suspected of being present and the coexistence of different conformational states in the gas phase can be detected by FTICR- MS. Under these conditions, cytochrome c was found to exist in multiple conformational states that were stable for hours, whereas in solution, H/D exchange involves nearly all of the exchangeable hydrogens, indicating rapid equilibration between conformers. By MS/MS, it has been shown that terminal ci-helices that are most stable in solution are not favoured in the gas phase. Conversion between states can be achieved by infrared laser heating, which resembles thermal unfolding in solution, by high-velocity collisions, which completely denature proteins, or by charge stripping with a proton-affinity agent to mimic solution pH increases and promote fewer open states.

Despite ion mobility and collision cross-section studies that show strong correlation with the charge state and reflect open conformations, it is surprising that the rates of H/D exchange are at best not affected. It is expected that a systematic increase in the exposed amide sites would lead to an increase in the numbers of amide hydrogens exchanged when exposed to $D_2O$ in the gas phase. In fact, an increase in the number of charges actually results in less H/D exchange. A number of explanations have been offered such as local chemical environments and energetically favourable hydrogen bonding that protect amide sites even in diffuse proteins. Many peptide ions and even energetic metastable ions retain their secondary structures on the microsecond time scale, which serves to protect the hydrogens involved in hydrogen bonding for the duration of the mass spectrometric experiment. It is still more difficult to rationalize the linear decrease in the numbers of exchangeable amide hydrogens with increasing charge. A possible explana tion is that the extra charges appearing on the side chains become solvated and, therefore, do not affect exchange.

# 6

# Dissociation of Peptides and Proteins

Evidence for the gas-phase CID of *molecular ions* is apparent in the first mass spectra recorded by Sir J. J. Thomson with his *parabola* mass spectrograph, and the phenomenon was a subject of study throughout the development of MS during the first half of the twentieth century. A summary of the early work on CID has been published. The modern application of CID to the detection, identification, and structural analysis of *organic molecules*, to *complex mixture* analysis, and to biopolymer sequencing can be traced to multiples works on CID itself and on the closely allied topic of *metastable* ion dissociation. Detailed study of CID during the latter half of the twentieth century has resulted in an understanding of the energy transfer *mechanisms* and dissociation chemistry of small ions (<500 Da). Tandem mass spectrometers, which use gas-phase collision regions to produce product ions from precursor ions, have proven extremely useful for the identification and characterisation of ions and for complex mixture analysis. This usefulness is due to several factors, including the ease of implementation of CID (relative to alternative methods such as photo- or surface-induced dissociation [SID]), the fact that CID is universal (i.e., all ions have a collision cross section), and the fact that CID cross sections are typically high.

The last 20 years has seen a revolution in the application of MS and tandem MS to *biological problems*, due to the advent of ionisation methods such as matrix-assisted laser *desorption* ionisation (MALDI) and electrospray ionisation (ESI), capable of producing ions from *biologically* derived samples such as peptides, proteins, and nucleic acids. Given the popularity and wide application of CID for use in tandem MS of small ions, it is not surprising that CID has been extended to the tandem MS of these larger biological ions. Indeed, tandem MS with CID has become an important tool in the growing field of proteomics, the effort to elucidate the protein complement for a given species as a function of *physiological* conditions. As is often the case in science, the application of the technique has outpaced the understanding of the underlying *phenomena* that dictate the appearance of the resulting data. The energy-transfer *mechanisms* that operate during collisions of large (>1000 Da) possibly multiply charged ions with small target gas atoms or *molecules* are not as well understood as they are for smaller ions. The dissociation chemistry of large multiply charged biological ions upon activation by collisions is also the subject of continuing study. The intent of this chapter is to describe the important experimental variables that affect the CID of biologically derived *macro-ions* and to describe what is known about how such ions behave as a function of those variables, with reference to examples from the *literature*. The discussion is largely focused on the behaviour of peptide and protein ions, as these species constitute by far the most widely studied biological ions by CID. The ranges of the relevant variables that can be accessed with available instrumentation is discussed. Note that in terms of instrumentation, the discussion is kept narrowly focused on CID behaviour; other characteristics of tandem mass *spectrometers*, such as precursor and product mass resolution, efficiency of ion transfer between mass analysis stages, sensitivity, and others, are not discussed. The reader should be aware that although a given instrument may access a range of CID variables, which yields useful tandem mass spectra in terms of the dissociation that occurs, other instrument criteria must also be considered when evaluating the applicability of a tandem mass spectrometer to a particular problem.

## Experimental Variables that Influence the CID Behaviour of Biological Ions

A wide variety of conditions has been used to effect CID of biological ions as new types of tandem mass spectrometers have been developed or as existing types have been fitted with ion sources capable of producing biological ions. Some activation methods have been developed specifically to improve the quality of tandem MS data for biological ions. A set of figures of merit for CID are outlined here to aid in the discussion of how *peptides* and *proteins* dissociate as a function of the CID conditions used:

1. The time over which the activation occurs relative to the time for *unimolecular* dissociation or rearrangement.
2. The amount of energy that can be deposited during the activation.
3. The distribution of the deposited energy.
4. The variability of the deposited energy.
5. The form of the deposited energy (*vibrational* vs. *electronic*).
6. The time scale of the instrument used (the kinetic window within which dissociation reactions must occur in order to be observed).
7. The efficiency of the CID process, in terms of cross section or rate constant.

The first figure of merit, the time scale of the activation, refers to how fast energy is deposited in the ion relative to how fast energy *equilibrates* through the degrees of freedom of the ion, how fast the ion dissociates or rearranges, and how fast energy is removed by cooling processes such as deactivating *collisions* or *photon emission*. McLuckey and Goeringer have distinguished three time regimens for activation methods used in tandem MS, including collisional methods: fast, slow, and very slow (slow heating). Figure elsewhere in this chapter illustrates typical ranges of activation times for a number of activation methods, with the collision based methods of interest here highlighted.

Fast methods are those in which the input of energy occurs rapidly relative to *unimolecular* dissociation, such as single collision CID at kilo-electron volts or electron volts (in the laboratory frame of reference) *kinetic energy*. Slow CID methods are those in which multiple collisions can occur, with long (tens to hundreds of microseconds) intervals between individual *collisions*. In this regimen, chemistry can occur between collisions, meaning that ions can dissociate or isomerize during the activation. This type of CID is effected in *collision cells*, such as the *central collision quadrupole* of *triple quadrupole* tandem mass spectrometers and other instruments that use multipole collision regions, which are operated at pressures of a few millitorr to as high as a torr Dissociation of Peptides and Proteins. Very slow activation methods, or slow heating methods, are distinguished from slow methods by the fact that deactivating processes, such as cooling collisions or photon emission, can occur between activation events. In this regimen, a steady state between activation and

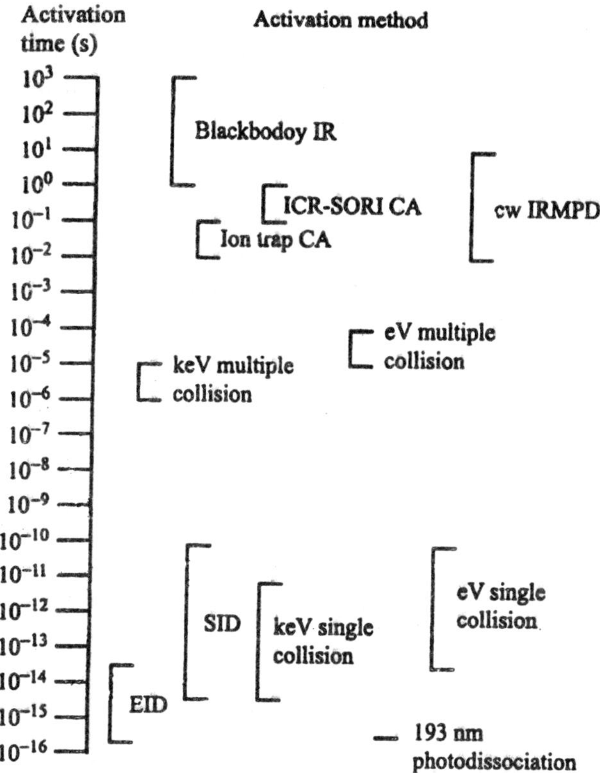


*Figure 6.1 : Activation time scale (time between activating events, e.g., gas phase collision or photon absorption) for a wide variety of methods used to dissociate biological ions, with the gas phase collision-based methods highlighted in bold.*

deactivation can be achieved. If the dissociation rate is low relative to the activation/deactivation rate (i.e., if the ions are in the so-called "*rapid energy exchange*" condition), then the ion internal energy can be described with Boltzmann statistics. To a good approximation, the ions can be considered to have an internal temperature, and ion dissociation can be described by the Arrhenius theory. Quadrupole ion trap collisional activation in approximately 1 *millitorr* of *helium* bath gas and sustained off-resonance irradiation (S ORI) in Fourier transform–ion *cyclotron* resonance (FT-ICR) MS are examples of slow heating CID methods. Direct heating of the helium bath gas is another form of slow heating that can be used for CID in quadrupole ion traps. Multiple excitation collisional activation (MECA) and very low-energy collisional activation (VLE-CA) are other forms of slow heating that have been used in FT-ICR. Slow heating methods lead to similar protein and peptide dissociation as slow methods, although an important difference is that for slow heating methods, the product ions are generally not subject to further activation, and hence less ***sequential*** dissociation is observed.

The amount of energy that can be deposited in an ion depends on the relative collision energy of the colliding ion/neutral pair. Most CID experiments can be roughly categorized into one of two energy regimens: "***high-energy***" CID with kiloelectron volt kinetic energy in the laboratory frame of reference and "*low-energy*" CID with 1-100 eV *kinetic energy* in the laboratory frame. It bears mentioning here that as discussed throughout this section, there are many variables that determine the amount, distribution, form, and so on, of internal energy deposited into the ion, and further, there are a number of variables in addition to those pertaining to the internal energy that determine the appearance of a CID spectrum. Hence, the historic distinction between "*high*"- and "*low*"-energy CID based only on the laboratory kinetic energy is an oversimpliûcation that should probably be avoided. Some statement regarding the collision energy, together with an indication of the number of *collisions*, the mass of the collision partner, and the activation time is probably a minimum in describing a particular CID experiment.

The kinematics for collisions of *small ions* with neutrals have been carefully described and show that it is the relative kinetic energy of the collision partners that is available for transfer to internal energy of the ion. This center-ofmass collision energy, $KE_{com,}$ is a fraction of the laboratory kinetic energy, $KE_{tab}$, as given by Eq. (1) if the velocity of the *neutral* is ignored:

$$KE_{com} = \left(\frac{N}{m_p + N}\right) KE_{tab}, \qquad (1)$$

where N is the mass of the neutral target and $m_p$ is the mass of the ion. Equation (1) shows that the energy available for transfer into internal modes for a collision of a high m/z ion with a small neutral is very low and suggests that CID of high m/z ions should be very inefficient. However, Maržluff et al. have shown that at least for low center-of-mass kinetic energy, the energy transfer process in collisions of large ions with small targets can be extremely efficient. In this study, a homologous series of *deprotonated peptides*, which all dissociate via the same well-characterized mechanism, was used to study the efficiency of dissociation as a function of peptide size.

The results indicate that the extent of dissociation of these peptides upon SORI-CID in an FT-ICR using nitrogen collision gas increased with the size of the peptide. RiceRamsperger-Kassel-Marcus (RRKM) statistical dissociation rate calculations for one of the peptides studied, gly-gly-ile, show that at the observed dissociation rate, approximately 55% of the available kinetic energy was transferred to internal energy during each collision. Trajectory calculations conducted with a molecular mechanics force field support this experimental observation and indicate that for a larger system such as bradykinin (nine *amino acid* residues and 444 internal degrees of freedom), as much as 90% of the available kinetic energy could be transferred.

Douglas has developed an approximation for the internal energy deposited into large ions upon electron volt collisions in a linear quadrupole, which supports the observations described earlier. Using experimental dissociation cross sections (see discussion of efficiency below) measured as a function of center-of-mass collision energy, Douglas showed that electron volt collisions of *bromobenzene* cations with *nitrogen* lead to approximately 70% conversion of center-of-mass energy to internal energy. Further work by Douglas aimed primarily at measuring collision cross sections for protein ions in a *triple-quadrupole* instrument provided further evidence for the efficiency of energy transfer. The collisions of large ions with small target gas atoms were shown to be most accurately modeled with a diffuse scattering model, wherein a small target atom collides with the ion at the center-of-mass collision energy, and then leaves with a thermal energy distribution; the result is that 90–95% of the collision energy is converted to internal energy of the ion during the relatively long interaction between the ion and the neutral. Experimental measurement of the dissociation threshold for the loss of heme from holomyoglobin supported this large energy transfer of more than 90%.

A number of authors have developed an impulsive collision theory (ICT) to describe the energy transfer process for collisions of large ions with small neutrals. In this model, direct momentum transfer between a constituent atom of the ion and the collision partner occurs, resulting in much larger transfer of internal energy to the ion than would be *predicted* based only on consideration of the overall mass of the large ion (Eq. [1]). Such binary collisions between portions of a large ion and the small target have been examined by other researchers. Bradley et al. compared the energy transfer to large *organic ions* consisting of many small atoms (C, N, O, H) with that to ions of similar m/z consisting of large atoms (CsI clusters) and showed that the behaviour was consistent with the prediction of the ICT, viz. more energy transfer to the *organic ion* during collisions with helium, due to the closer similarity in mass between the small atoms of the organic ion and the helium atoms. The reverse was found to be true for collisions with *argon*, because of a closer mass match between the Cs and I atoms of the cluster and the argon target atoms.

Turecek has shown that in principle, the center of mass limit on the amount of energy deposited can be overcome by invalidating the assumption made in Eq. (1), viz. that the velocity of the neutral collision partner is zero. By colliding ions with fast (*kiloelectron volts*) rather than thermal targets, much more energy is available for transfer to internal energy during the collision. However, the practical difficulties of crossing a fast beam of ions with a fast beam of neutrals or of *colliding* a fast beam of neutrals with a small stationary ion cloud, with sufficient cross section to yield measurable product signals may limit the analytical applicability of this form of CID. Note that crossed beams have been used for fundamental studies of ion–molecule reactions at very low *kinetic energy*.

The number of collisions occurring during the activation period also determines the amount of energy deposited. Most analytical applications of CID use collision regions operated at pressures high enough to ensure that multiple collisions occur to increase internal energy and hence yield extensive dissociation. The disadvantage to using higher pressure is that increased *scattering* may result in a loss of sensitivity; however, Douglas and French have shown that in quadrupole *transmission* devices used to carry ions from an ESI source to the mass analyzer, collisional cooling of ions at elevated pressure (~8 millitorr) leads to improved sensitivity and resolution (by removal of kinetic energy spread in the ions). This concept has also been applied to quadrupole collision cells of tandem mass spectrometers, leading to improved product ion collection efficiency and resolution when the cells are operated at 8 millitorr compared to the 1–2 millitorr normally used.

This discussion relates to the magnitude of the energy that can be deposited; the third and fourth figures of merit relate to the distribution of energies that can be deposited and the variability of the energy distribution. The distribution of energies plays an important role in determining the appearance of CID spectra. Ideally, the energy distribution is narrow, well defined, and variable over a wide energy range, so dissociation channels having a wide range of critical energies can be accessed. CID methods lead to energy distributions that are typically fairly broad and ill defined, at least for fast and slow CID methods. In kiloelectron volt collisions, a distribution having a maximum at only a few *electron volts*, with a long tail out to higher energy (tens of electron volts), is often obtained for small ions, allowing some high critical energy channels to be observed. For electron volt collisions, the maximum of the distribution is similar, but the high energy tail is not present, so that only lower critical energy processes are accessed. Douglas has presented a method for estimating the internal energy distribution for multiple collision CID in a quadrupole from a measure of the dissociation cross section as a function of collision energy. For ion trap collisional activation, it has been shown that the internal energy distribution can be modeled as a Boltzmann or truncated Boltzmann distribution, depending on the relative rates of activation/deactivation and dissociation. Schnier et al. have shown that for FT ICR SORI, leucine enkephalin and bradykinin ions achieve effective temperatures of between 500 and 700 K, although the authors state that actual distribution of internal energy is not well characterized and may not be Boltzmann. As discussed earlier, the maximum of the internal energy distribution can be varied to some extent by varying the collision energy; CID spectra acquired at low laboratory kinetic energies (electron volts) typically show greater sensitivity to changes in energy than spectra acquired at *kiloelectron volt* energy.

The fifth figure of merit relates to whether the internal energy is present in the activated ion in electronic or vibrational modes. *Statistical theories* of activated ion dissociation assume that internal energy equilibration through all *vibrational degrees* of freedom is faster than dissociation. Experimental evidence shows that this holds for most small *organic ions*; however, exceptions where dissociation from an *electronic* excited state is faster than relaxation of energy into vibrational modes have been reported. There is little evidence for or against the *statistical hypothesis* for large peptide and protein ions. However, it seems reasonable, at least under multiple collision conditions where energy is deposited in *small increments* at a variety of *collision sites* on the ion, that collisional activation effectively gives rise to a statistical ion energy distribution.

The sixth figure of merit relates to how fast the dissociation reactions must occur in order to be observed on the time scale of the instrument used. The kinetic window varies widely across different instrument types from a few tens to hundreds of microseconds for magnetic sector-based instruments operating at kioelectron volt ion kinetic energies, milliseconds in triple quadrupole and quadrupole TOF instruments, and from hundreds of milliseconds up to seconds in trapping instruments. The length of the kinetic window is important for large biological ions, because dissociation of these ions may be relatively slow because of their large number of degrees of freedom. The amount of energy above the critical energy for dissociation required to drive a dissociation reaction at a rate that is observable on the time scale of the instrument used is called the kinetic shift and may be very large (up to or even exceeding 100 eV) for large biological ions. This obviously makes the previous discussion regarding the amount of energy that can be deposited into the ions upon collisional activation very important (e.g., at kioelectron volt kinetic energies, dissociation rates must be driven to at least $10^4$ $s^{-1}$ in order to be observed), and hence inherently slow processes may not be observable in this regimen, due to an inaccessibly high kinetic shift.

The efficiency of the CID process relates to how much of the available precursor ion is converted to product ions. For beam type instruments, such as sectors and *triple quadrupoles*, this is most conveniently represented with a Beer's law type of relationship, as shown in Eq. (2):

$$\left[M_p^+\right] = \left[M_p^+\right]_0 e^{-n\sigma l}, \qquad (2)$$

where $\left[M_p^+\right]_0$ is the precursor ion flux without collision gas, $\left[M_p^+\right]_t$ is the precursor ion flux after the introduction of *collision gas*, 1 is the path length, n is the target number density, and a is the total ion loss cross section. If no processes for precursor ion loss other than CID operate, then a represents the cross section for CID. For trapping instruments, an *analogous* expression in terms of rate constant and reaction times can be used:

$$\left[M_p^+\right]_t = \left[M_p^+\right]_0 e^{-nkt}, \qquad (3)$$

where $\left[M_p^+\right]_0$ is the precursor ion abundance before any reaction, $\left[M_p^+\right]_t$ is the precursor ion abundance after reaction time t, and k is the rate constant for all precursor ion loss processes. If no precursor ion loss processes other the CID operate, then k represents the rate constant for CID. Equation (2) shows that the only variable available to increase the efficiency of CID in beam-type instruments is n, the number density of the neutral collision partner (increasing the path length is possible but is usually not straightforward over a wide range). The advantages and disadvantages of operating at higher collision gas pressure were discussed earlier. At higher pressure, ion loss processes such as scattering may degrade sensitivity, although in quadrupole collision cells, the collisional cooling afforded by higher collision gas pressure has been shown to give the opposite effect. It is easier to increase the efficiency of CID in ion-trapping instruments by increasing the reaction time, t. Reaction times as long as seconds can be used to completely drive precursor ions to product ions. Note that operation at either higher *collision gas* pressure or longer activation time increases the likelihood of multiple collisions occurring and, hence, implies that the activation is either slow or very slow (slow heating), with the attendant possibility for chemistry to occur during the activation, as discussed earlier.

**Summary of Commonly Used CID Conditions**

The values for any individual figure of merit discussed earlier may vary widely with the type of instrument used and the design of the experiment, and the various figures are not independent but influence one another. In this section, three commonly used regimens for effecting CID of gas-phase *biological ions* are described in terms of the figures of *merit* outlined earlier. The information given in this section with typical ranges of values used for the operating parameters and figures of merit for the three regimens. CID spectra for a common ion, the doubly protonated ion of the nine residue peptide *bradykinin* (RPPGFSPFR), are used throughout this section to illustrate the influence of the figures of merit on CID behaviour.

***High-Energy CID (Fast Activation)***

As discussed earlier, the collision energy is only one of a number of variables that must be specified to adequately describe a CID regimen. The term "*high-energy*" CID is usually used to describe CID effected at kiloelectron-volt precursor ion kinetic energies, with a target gas pressure that is low enough that only single or at most a few (fewer than five) collisions can occur. Because of the high kinetic energy of the ions, the time scale for dissociation is usually on the order of a few *microseconds*. The vast majority of CID experiments under this regimen have been carried out using sector-based *instrumentation*. However, *kioelectron-volt* collisions are seeing increasing use in tandem time-of-flight (TOF) instruments, as discussed later in this chapter.

Double-focusing mass spectrometers based on *magnetic* and *electric* sectors were the first instruments used for CID studies of small organic ions, and were among the first used for CID of peptide ions approximately a decade later. Two-sector MS instruments may be used to record MS/MS spectra by using a variety of linked scans of the *magnetic* (B) and *electric* (E) fields, albeit with relatively poor precursor ion or product ion resolution, depending on the type of scan. A number of three-sector instruments were constructed to help overcome this limitation. Four-sector instruments, true tandem mass spectrometers that combine two double focusing instruments for high resolution of precursor and product ions, were also constructed and were commercially available from a number of manufacturers. The use of these instruments for tandem MS of *biological ions* was demonstrated throughout the 1980s and early 1990s, but they have now been largely superseded by tandem quadrupoles, hybrid quadrupole/TOF instruments, ion-trapping instruments, and increasingly TOF/TOF instruments.

Readily obtainable field strengths for sectors are such that these instruments must operate on ion beams accelerated to kiloelectron-volt *kinetic energies*; hence, most CID work with sector instruments is done with collisions in this regimen. As discussed earlier, kiloelectron-volt collisions deposit on average a few electron volts of *internal energy*, but the distribution can include a tail to higher internal energies, so that higher critical energy or lower rate processes (requiring large *kinetic* shifts) can be observed. High-energy CID spectra are relatively insensitive to changes in the kinetic energy of the precursor ion, and so obtaining energy-resolved CID spectra with high-energy CID is difficult. However, it has been shown that higher energy collisions lead to larger scattering angles, so recording MS/MS data as a function of angle (angle-resolved MS [ARMS]) yields information on dissociation as a function of internal energy. It is possible to decelerate and then reaccelerate the beam to also access low-energy collisions; however, high-energy CID spectra are very *reproducible* on a given instrument and even between instruments of different configuration, and this is often *cited* as a major advantage so that operation at low energy is usually not considered to be a desirable alternative.

The transit time of a 10-keV beam of ions at m/z 1000 through a 10-cm collision cell is 2.3 ms, and 23 ms along a 1-m ûight path from the collision cell to the *detector*. The former value establishes the maximum amount of time between activating collisions, if the cell is operated at a pressure at which multiple collisions in this time frame are likely. High-energy CID is normally performed using pressures chosen to yield a given *attenuation* of the ion beam, which is related to the number of collisions. Single or at most a few (5–10) collisions are typically used; beyond this, *scattering* of the beam leads to unacceptable losses in sensitivity. Increasing the number of *collisions*, of course, increases the energy deposited. At least for small ions, internal rearrangements can occur on a time scale of nanoseconds or less, hence isomerisation can occur between *collisions* even at high kinetic energy. The transit time through the whole instrument of some tens of *microseconds* establishes a lower limit on the observable rates of approximately $10^4$ to $10^6$ $s^{-1}$.

For peptides, the figures of merit described lead to CID *spectra*, which differ from those obtained using other commonly employed conditions primarily in that abundant dissociation of *amino acid* side chains is observed in addition to peptide bond cleavage. The MS/MS spectrum of +2 *bradykinin* obtained on a four-sector instrument using 8-keV collisions with *helium* at a pressure at which the primary beam intensity reduced by 70%, corresponding to between two and four collisions. The ions labeled wn and dn correspond to side-chain cleavage. It has been suggested that these cleavages are charge remote and, hence, require higher critical energies to be accessed, implicating the high-energy tail of the internal energy distribution as the cause for these cleavages. This is supported by the absence of side-chain cleavages in CID spectra collected at *electron-volt* collision energy (see later discussion). These side-chain cleavages have been shown to be useful for the distinction of isomeric and isobaric amino acids in peptide-sequencing applications. The upper limit for CID at high energy is approximately m/z 3000, beyond which dissociation is not readily observed, probably because of the large kinetic shift required to drive dissociations at a rate of at least $10^4$ $s^{-1}$. The efficiency of high-energy CID is typically only a few percent even for smaller molecules because of this short time scale and because of the product ion collection efficiency.

The advantages of CID conducted at kiloelectron-volt energy and low collision numbers, viz. reproducibility and access to side-chain cleavages for isomeric and isobaric ion distinction, are driving interest in an alternative instrument, the tandem TOF (TOF/TOF), for accessing this regimen. An advantage of the TOF/TOF over sector-based instruments is the well-established compatibility of TOF with the MALDI source. A number of forms of TOF/TOF have been described for CID and photodissociation of small molecules and cluster ions. At least two groups have demonstrated the use of TOF/TOF instruments to dissociate peptide ions with kiloelectron-volt cóllisions. A CID spectrum for a tryptic peptide dissociated using 3-keV collisions with argon (pressure unspecified). The abundant immonium ions of individual amino acids and the wn ions resulting from side-chain cleavage are characteristic of the high collision energy used and were helpful in establishing the peptide composition and distinguishing the isomeric leucine and isoleucine residues, respectively.

***Low-Energy CID (Slow Activation)***

The term ''*low-energy*'' CID is typically used to refer to CID conducted in quadrupole or other multipole collisions cells and is characterized by collision energy of up to 100 eV, target gas pressures selected to allow multiple (tens to hundreds) collisions, and a time scale on the order of a few hundred microseconds to a few milliseconds. For simplicity, throughout this discussion the term quadrupole, represented with the letter q, will be used to describe a multipole collision cell that passes all m/z ratios above a certain lower limit, with the understanding that higher multipoles such as *hexapoles* and *octapoles* are in some cases used instead of quadrupoles. The electron-volt energy regimen is used in tandem quadrupole instruments (usually referred to as triple quadrupoles [QqQ], where Q is a mass-resolving quadrupole) and in hybrid instruments that combine quadrupole collision cells with other types of mass analyzers for precursor selection and/or product mass analysis. A hybrid instrument that is increasing in popularity is the quadrupole-quadrupole-TOF (QqTOF), where precursor selection is done with the first quadrupole, CID occurs in the second quadrupole, and the products are mass analyzed via TOF. The attractive mass analysis characteristics of orthogonal acceleration, reflectron TOF are driving interest in this instrument, and two commercial versions are available.

The advantage of using a radiofrequency (RF) quadrupole (or other multipole) as a collision cell is that the RF field provides a strong focusing force toward the ion optical axis, so that losses of precursor and product ions are minimized, even at relatively high collision numbers. However, to enjoy this advantage, ions must move through the quadrupole slowly enough to be influenced by the rapidly changing RF field; hence, kinetic energies less than 100 eV are typically used. For small organic ions, low-energy collisions lead to internal energy distributions that have a peak at a few electron volts of internal energy, but without the long tail to higher energy, which characterizes kioelectron-volt CID. The observed differences between electron-volt and kiloelectron-volt dissociation of peptides is that electron-volt CID lacks the side-chain cleavage peaks often observed at higher energy. The MS/MS spectrum of +2 *bradykinin* collected on a triple-quadrupole instrument using electron-volt collisions (between 70 and 200 eV) with argon at a pressure that yielded a target-gas thickness of 2–5 $\times 10^{14}$ atoms/cm$^2$, where target-gas thickness is equal to the product of the number density of the gas and the length of the collision cell. The absence of w- and d-type ions in this electron-volt kinetic energy spectrum supports the suggestion that side-chain cleavages leading to these ions are high-energy processes. Some sequential dissociation, leading to internal fragments (e.g., $y_7b_{5)}$ and immonium ions (P and F), is observed.

As discussed earlier in this chapter, Douglas has estimated that collisions of large ions with small targets can be very efficient for the transfer of available kinetic energy to internal energy; therefore, provided that sufficient collisions occur, even large proteins having many degrees of freedom, and hence large kinetic shifts can be driven to dissociate at observable rates. This is illustrated in the data in Fig. 5, which shows the dissociation of the (M + 20H)$^{+20}$ ion of *apomyoglobin* collected on a *triple-quadrupole* instrument at 2-keV collision energy using collisions with argon at a pressure that yielded a target-gas thickness of 1 $\times$ $10^{14}$ atoms/cm$^2$.

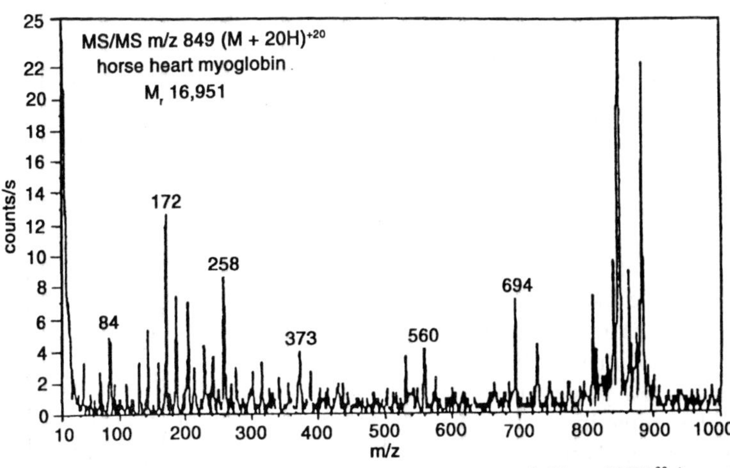


*Figure 6.2 : Collision-induced dissociation (CID) spectrum of (M + 20H)$^{+20}$ ions of horse heart apomyoglobin collected on a triple-quadrupole instrument at 2-keV collision energy using collisions with argon at a pressure that yielded a target-gas thickness of 1 $\times$ $10^{14}$ atoms/cm$^2$.*

Note that this experiment is at the interface between the low-energy and high-energy regimens, highlighting the danger of describing an experiment based only on the collision energy. Because the force acting on an ion is proportional to the charge on the ion, a higher collision energy can be used for this highly charged protein ion while still enjoying the focusing advantages of the linear quadrupole discussed earlier in this chapter. At the pressure used in this study, multiple collisions were likely, so the experiment was a slow activation experiment with ample time for rearrangement before dissociation. Dissociation of this large precursor was aided by the fact that the ions were already "*heated*" in the electrospray interface by use of large potential differences between the lenses used to transport ions into the mass spectrometer.

CID spectra collected at electron-volt collision energy can be more sensitive to changes in the collision energy than CID at kiloelectron-volt energy, allowing the effects of internal energy on ion dissociation to be probed in this regimen. This technique is referred to as energy-resolved MS (ERMS). ERMS is used to generate break down graphs, or plots of the relative contributions of ion dissociation channels as a function of ion *internal energy*. Such graphs can be useful for elucidating reaction *mechanisms* and for distinguishing *isomeric species*.

At 100 eV, an m/z 1000 ion will travel through a 20-cm *quadrupole* in approximately 1.5 ms. This time scale means that extensive isomerisation of the ion of interest may occur between collisions if multiple collision conditions are used. Processes with rate constants of 103 $s^{1}$ or greater can be observed in this relatively slow CID regimen. Collision gas pressures of 1–2 millitorr, leading to some tens of collisions, have traditionally been used to effect CID in a quadrupole. However, a major innovation in the use of quadrupoles for CID of larger *biological ions* is the demonstration that collisional damping of radial motion in the quadrupole allows higher pressures, and hence more collisions, to be used without excessive ion losses due to scattering. Operation at approximately 8 millitorr has been shown to improve the efficiency of triple-quadrupole MS/MS from only a few percent to 30–50%. Note that much of this improvement is due to the improved collection and transmission of product ions by the mass analyzing quadrupole (Q3) after the collision cell due to the removal of the spatial and kinetic energy spread of the products. At higher pressure, transit times through *collision cell* can be very long

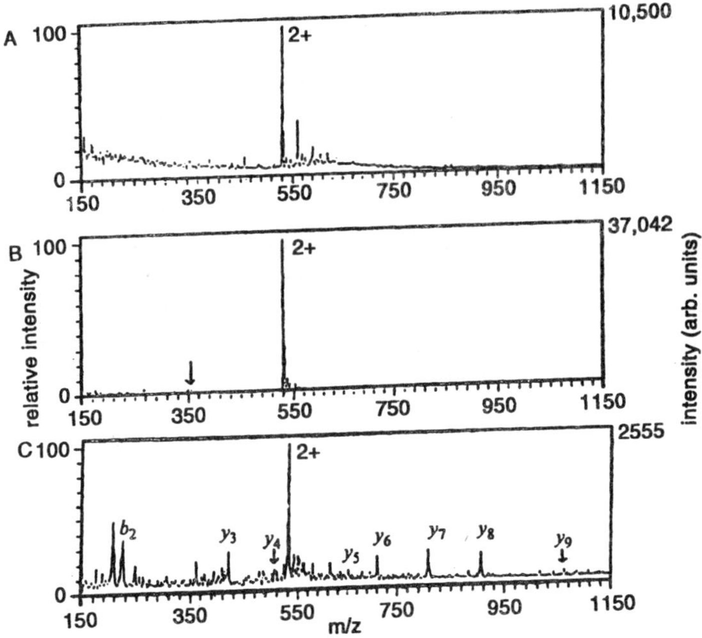


*Figure 6.3 : Spectra of bradykinin recorded as a function of the potential difference between interface lenses of the electrospray source. (A) The potential difference was 80 V, and only the $(M + 2H)^{+2}$ ions and $(M + 2H)^{+2}$ with solvent molecules attached were observed. (B) The potential difference was increased to 160 V, so that the $(M + 2H)^{+2}$ was efficiently desolvated. (C) The potential difference was further increased to 260 V, and abundant b- and y-type product ions were observed due to nozzle/skimmer collision-induced dissociation (CID).*

(tens of milliseconds) unless an additional electric field in the direction of ion motion through the quadrupole is added to "*pull*" the ions through the gas. When such a field is added, pressures of tens to hundreds of millitorr and as high as 3 torr may be used to effect CID. Under these conditions, dissociation reactions with rates as low as a few tens per second can be accessed.

Another form of CID conducted at electron-volt collision energy is the so-called "*in-source*," "*cone-voltage*," or "*nozzle/skimmer*" CID effected in the interface region of ESI sources. The lens voltages that guide ions through the intermediate pressure region between the *atmospheric pressure* source and the *high vacuum* of the mass analyzer region may be manipulated to cause ions to undergo collisions with residual background gas at *collision energies* up to a few hundred electron volts. The time scale for this type of dissociation is on the order of a few hundred microseconds to a few *milliseconds*. Although this is not a true MS/MS experiment because there is no prior selection of the precursor ion, nozzle/skimmer CID can provide efficient dissociation of peptides and proteins on simple instruments. For example, the nozzle/skimmer CID spectrum of +2 bradykinin collected at 520 eV in the approximately 1–10 torr region of an ESI source interfaced to a single quadrupole is shown in Figure elsewhere in this chapter (bottom panel). The spectrum is similar to that collected via true MS/MS with a triple-quadrupole instrument, in that only b- and y-type ions are observed. Nozzle/skimmer dissociation has also been demonstrated for whole protein ions. Figure elsewhere in this chapter shows the dissociation of multiply charged carbonic anhydrase B cations in an electrospray source interfaced to an FT-ICR instrument. Nozzle/skimmer CID has also been used for ERMS of small peptides, with comparable results to those obtained for quadrupole ERMS.

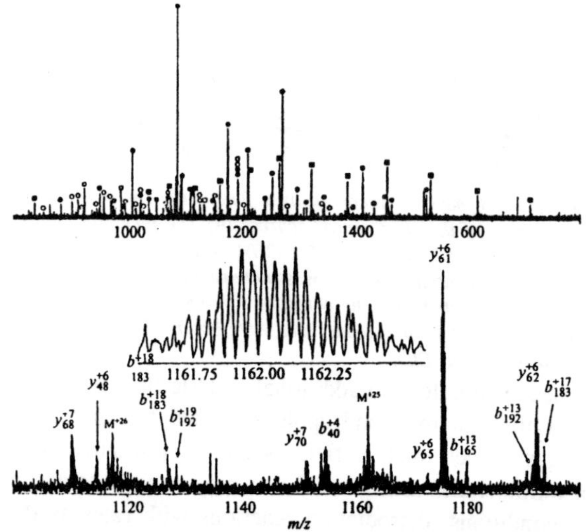


*Figure 6.4 : Nozzle/skimmer dissociation of multiply charged carbonic anhydrase B cations in an electrospray source interfaced to an Fourier transform-ion cyclotron resonance (FT-ICR) instrument, with a potential difference of 165 V across the ion source.*

### *Ion-Trapping CID (Very Slow Activation, Slow Heating)*

The two forms of ion traps used in MS, *electrodynamic* quadrupole ion traps and *electrostatic/magnetic* ICR ion traps, vary in the mechanism by which they store ions and perform mass analysis; however, their behaviour with respect to CID is quite similar. CID in both instruments is characterized by many collisions (hundreds) at a few electron-volt (tens) collision energy and, as such, is qualitatively similar to the "*low-energy*" quadrupole CID discussed in the previous section. An important difference is that the trapping nature of these instruments means that very long time scales can be used to achieve very high dissociation efficiency (75–100%) and to access very slow dissociation channels.

Another consequence of the long time scale accessible with trapping instruments is that deactivating processes can also occur, so that a steady state internal energy can be achieved. Details of CID in a quadrupole ion trap are discussed first in this section, followed by discussion of CID in an FT ICR trap.

*Quadrupole Ion Trap CID*

Quadrupole ion traps store ions in the rapidly changing potential well generated by the application of an RF voltage to the trap electrodes. Ion motion in the field depends, inter alia, on the frequency and amplitude of the RF voltage and on ion m/z. For a given RF frequency and amplitude, a selected precursor m/z will have a characteristic frequency of motion in the trap, normally referred to as the secular frequency. Application of a supplementary voltage matching this secular frequency will excite the precursor ions, causing them to undergo energetic collisions with background gas or deliberately added collision gas (usually helium, see below), thereby increasing their internal energy. This process, commonly called resonance excitation, was first demonstrated by Louris et al. in 1987. The depth of the trapping potential well is such that the ions can only be excited to a few electron volts of kinetic energy before they are ejected from the trapping volume by the supplementary resonant voltage. For this reason, multiple collisions over a relatively long time (tens to hundreds of milliseconds) are required to build up sufficient internal energy to lead to dissociation. Clearly, extensive rearrangement of the precursor ion can occur during the activation and before dissociation. Another consequence of this long slow buildup of internal energy is that deactivating processes such as IR emission and/or cooling collisions can also occur. Goeringer and McLuckey have developed a collisional energy transfer model for resonance excitation, which shows that competition between activation and deactivation leads to a steady state internal energy that can be characterized by a Boltzmann distribution and an internal temperature, provided the dissociation rate is low compared to the activation/deactivation rate. If dissociation is rapid enough to deplete the high energy tail of the distribution, a truncated Boltzmann distribution results and the temperature is more correctly described as an "*effective*" internal temperature. For small peptides (e.g., leucine enkephalin, 556 Da), effective temperatures of approximately 650 K can be achieved. The available temperature decreases with increasing m/z as the difference between the mass of the ion and the collision partner increases. Another factor that limits the temperature to which high m/z ions can be raised is that for a given RF amplitude, higher m/z ions are stored in shallower potential wells and so must be excited with lower resonance voltage amplitudes to avoid ion losses through ejection from the trap. Current instrumentation limits the protein ion m/z on which CID can be performed to approximately 8600 Da/charge (+1 ubiquitin), although the upper limit for newer ion traps (e.g., the Finnigan LCQ, Bruker Esquire, or Hitachi M8000) has not yet been evaluated. Note that this limitation is stated in terms of mass/charge. Proteins with masses greater than ubiquitin can be readily dissociated with quadrupole ion traps as multiply charged ions with m/z lower that 8600 Da/charge; for example, a study of the dissociation of $(M + 2H)^{+2}$ to $(M + 21H)^{+21}$ ions of *apomyoglobin* (mass 16,950 Da) has appeared.

CID spectrum for +2 *bradykinin* acquired with 200 ms of resonance excitation in an ion trap having a background pressure of 1 millitorr of helium. Note that only b- and y-type ions are formed, in similar fashion to the electron-volt collision CID performed in a quadrupole collision cell and in an ESI interface. Whole-protein ions can also be dissociated via resonance excitation in a quadrupole ion trap. Figure elsewhere in this chapter shows the CID spectrum of $(M + 8H)^{+8}$ ion of ubiquitin. The precursor ion was excited for 300 ms in the presence of 1 millitorr of *helium*.

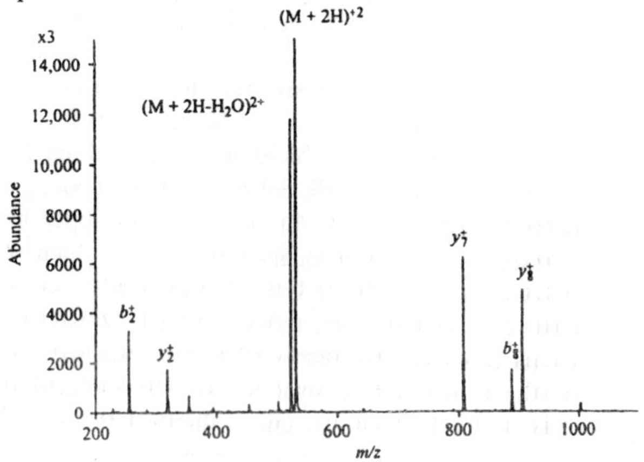

*Figure 6.5 : Collision-induced dissociation (CID) spectrum for +2 bradykinin acquired with 200 ms of resonance excitation in a quadrupole ion trap having a background pressure of 1 millitorr of helium.*

Although the internal temperature accessed during ion trap CID of this large ion is relatively low, the trapping nature of the instrument allows for long reaction times and hence low reaction rates, so relatively efficient dissociation is observed. An important point about quadrupole ion trap CID is that all product ions below a certain low mass cutoff (LMCO), determined by the RF amplitude, are not trapped. For CID of large ions, it is desirable to employ a relatively high RF *amplitude* to maximize the potential well depth and avoid ion losses by ejection; however, a high RF amplitude means that low m/z products are not trapped and so are not observed.

As discussed earlier in this chapter, ion trap CID is always carried out under multiple collision conditions, hence extensive rearrangement can occur during the activation period and before dissociation. The collision gas used is almost always helium, because helium bath gas is necessary to improve the resolution of ion trap mass analysis and so is already present in the system at a static background pressure of 1 millitorr. The effect of adding higher mass collision gas has been studied, and it has been shown that the deposited internal energy is increased, as predicted by Eq. (1). Several groups have reported that addition of a small fraction (5%) of, for example, argon, krypton, or xenon can improve the CID efficiency of peptide ions by allowing higher internal energy deposition. This has the effect of increasing the extent of dissociation and accessing higher critical energy dissociation channels. Addition of heavy gases also allows CID to be carried out at lower RF trapping voltages, which helps to overcome the LMCO limitation discussed earlier. Note that the upper m/z limit for CID in the ion trap given earlier (8600 Da/charge) was probably achieved at least in part because of the significant background pressure of air ($-1.5 \times 10^{-4}$ torr) present in the ion trap.

An alternative to raising the internal temperature of the ions in a quadrupole ion trap by increasing the kinetic energy of the ions via resonance excitation is to increase the kinetic energy of the neutral collision partner by heating the helium bath gas. Asano et al. have shown that ions reach thermal equilibrium with the bath gas, soif the temperature of the bath gas is known, and if the rate of activation/ deactivation is large relative to the rate of dissociation, then Arrhenius activation parameters can be derived from measurements of ion dissociation rate as a function of bath gas temperature. In addition, they have demonstrated the measurement of Arrhenius parameters for leucine enkephalin and bradykinin and have used leucine enkephalin as a thermometer ion to derive the temperature achieved via resonance excitation and boundary activated dissociation (BAD). Figure elsewhere in this chapter shows the CID spectrum obtained for +2 bradykinin at a bath gas temperature of 486 K and pressure of 1 millitorr, with a reaction time of 10 s. At this temperature, the facile cleavage at the N terminal of proline dominates, with loss of a water molecule and formation of one other complementary pair also observed at low abundance.

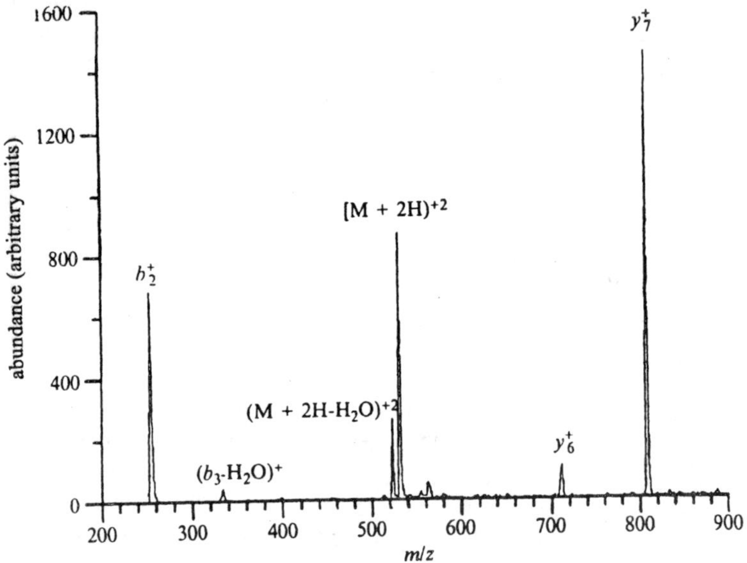


*Figure 6.6 : Collision induced dissociation (CID) spectrum for $(M + 2H)^{+2}$ bradykinin obtained by storing the ions in quadrupole ion trap at a helium bath gas temperature of 486 K and pressure of 1 millitorr for 10 s.*

A number of other alternatives for ion heating in quadrupole ion traps have also been demonstrated, such as direct current (DC) pulse activation (which yields a mixture of CID and SID products); low-frequency square-wave activation; irradiation with broadband waveforms such as filtered noise fields (FNFs), stored waveforms generated with inverse Fourier transforms (SWIFT), and random white noise; BAD; and off-resonance activation with a single frequency selected to be lower than the secular frequency of the precursor by approximately 5%. The last two techniques have been studied fairly extensively for their applicability to CID of peptide ions. The BAD technique involves applying a DC voltage to the ion trap electrodes to move ions close to the boundaries of stability in the trap, thereby increasing the amplitude of their oscillations and causing them to undergo energetic collisions with the bath gas. Glish and Vachet have shown that BAD can be used to dissociate peptides with the advantage that it is not necessary to tune the excitation to the exact frequency of ion oscillation to achieve maximum efficiency. However, the BAD technique suffers from relatively poor overall efficiencies due to the competition between ion ejection and dissociation. This limitation can be overcome somewhat by the combination of BAD with heavier collision gases. Asano et al. used leucine enkephalin to estimate that BAD can elevate ions to an effective internal temperature of approximately 700 K during collisions with helium, comparable to that obtained via normal resonance activation. Qin and Chait have shown that irradiating ions with a large-amplitude (21 $V_{pp)}$ alternating current (AC) voltage at a frequency approximately 5% below the secular frequency of the ion could more efficiently dissociate large peptides than normal on-resonance excitation. They attributed the increase in efficiency to the ability to deposit larger amounts of internal energy without ion ejection from the trap.

Fourier Transform–Ion Cyclotron Resonance CID. FT-ICR MS uses the combination of a high magnetic field (from 3 to as high as 20 Tesla) with small DC potentials to trap ions. Ion motion in the magnetic field exhibits a characteristic frequency, the cyclotron orbital frequency, which is m/z dependent; hence, a resonance excitation method analogous to that described earlier for quadrupole ion traps can be employed in FT-ICR traps to increase ion kinetic energy and thus cause internal energy deposition via energetic collisions. However, it has been shown that irradiating trapped ions at frequencies slightly (<1%) above or below the cyclotron frequency greatly improves the efficiency of CID in FT-ICR traps, because the ions can be held at elevated kinetic energy for long pe riods without ejection from the trap. This techni que is referred to as SORI. The mecha nism for SORI essentially involves elevatio n of the excited ions to a steady state kinetic energy after the first 10–20 ms of excitation due to dephasing of the ion motio n with respect to the applied excitation. Ion kinetic energies are comparable to those achieved via quadrupole ion trap resonance excitation (i.e., some few electron volts and hence many multiple collisions are required to increase ion internal energy enough to cause dissociation). The maximum kinetic energy accessible depends on the strength of the trapping magnetic field. Schnier et al. have used Arrhenius activation parameters measured via blackbody infrared radiative dissociation (BI RD) for leucine enkep halin and bradykinin ions to show that effe ctive temperatures of be tween 500 and 700 K are achieved during SORI using nitrogen as the collision gas. However, the authors are carefu l to state that the actual distribution of internal en ergy is not well charac terized and may not be Boltzmann. Laskin and Futrell have shown that smaller organic ions, such as bromonapht halene, achieve effective temperatures from 1300 to 4000 K during SORI. Preliminary attempts to use master equation modeling to develop a therma l model of SORI analogous to that developed for quadrupole ion trap CID have shown that the initial internal energy of the ions before activation and radia tive relaxation during activation both play significant roles in determining the ion internal temperature.

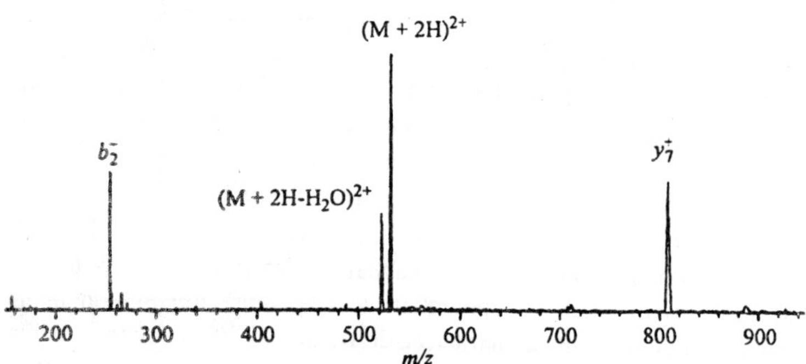


*Figure 6.7 : Collision induced dissociation (CID) spectrum of $(M + 2H)^{+2}$ bradykinin obtained via Fourier transform–ion cyclotron resonance (FT ICR) sustained off resonance irradiation (SORI) with irradiation for 500 ms at a nitrogen pressure of between 1 and $8 \times 10^{-6}$ torr. Note the similarity between this spectrum and those shown in Figs. 8 and 10 for quadrupole ion trap resonance excitation and bath gas heating to 486 K, respectively.*

The CID spectrum of +2 bradykinin obtained via SORI with irradiation for 500 ms at a nitrogen pressure of between 1 and $8 \times 10^{-6}$ torr. Note the similarity between this spectrum and quadrupole ion trap resonance excitation and bath gas heating to 486 K, respectively.

As discussed earlier in this chapter, Marzluff et al. has shown that transfer of available kinetic energy to internal energy is efficient at the kinetic energies accessed via SORI. Therefore, provided that sufficient collisions occur, even large protein ions can be made to dissociate. The SORI CID spectrum of $(M + 17H)^{+17}$ ion of apomyoglobin obtained by exciting the ions for 2 s in the presence of nitrogen (pressure unspecified). In analogy to quadrupole ion trap CID, the internal temperature of the ions is expected to be relatively low, but the trapping nature of the ICR instrument allows for very long excitation times, so processes with rates as low as 10/second or less can be observed.

The effects of the pressure and nature of the collision gas used in SORICID have not been widely studied. FT-ICR operates at very low ($10^{-9}$ torr) pressure during mass analysis to achieve ultrahigh resolution; therefore, collision gas for CID is pulsed into the cell. Because the collision gas is not present at a constant pressure during the CID event, characterisation of pressure effects is difficult. Gorshkov et al. have presented a model for estimating the average laboratory-frame kinetic energy, which accounts for the presence of collision gas, which can cool the ions if the pressure is too high. They have shown that there is an optimum gas pulse length and, hence, pressure peak during the CID event, which yields maximum internal energy deposition and CID efficiency. At shorter pulse times (lower pressure), an insufficient number of collisions occurs, and CID yields are reduced; at longer pulse times (higher pressure), collisional damping lowers ion kinetic energy, again limiting internal energy deposition and subsequent CID yields.

Two other methods for exciting ions for CID in an FT-ICR trap are also used, both with the same goal as SORI to increase ion kinetic energy without causing ion ejection from the trapping cell. MECA relies on a number (hundreds) of low-amplitude on-resonance excitation periods, each of which increases ion kinetic energy, but not to the extent that ions are ejected. Ion kinetic energy decreases between each activation period, but internal energy does not (at least not all the way back to the starting point), and so by using several excitation periods, internal energy can be increased until ions dissociate without causing ion injection. VLE-CA also uses on-resonance activation, but with rapid 180° phase shifts of the applied excitation voltage. These phase shifts prevent the applied voltage from increasing the ion kinetic energy to the ejection point. Senko etal. have evaluated the applicability of SORI, MECA, and VLE-CA (and the related resonant amplitude modulated collisional activation [RAM-CA]) to the dissociation of protein ions in an FT-ICR cell.

An important distinction between CID in both types of ion traps and the other types of CID discussed above is that because of the resonant nature of the excitation in ion traps, the productions are not themselves subject to further activation after their formation. Product ions are formed with the same internal energy as the precursor, on a "*per degree-of-freedom*" basis, and so they may still undergo further dissociation if the dissociation rate is faster than the rate of deactivation; however, they will not receive any further internal energy from the resonance excitation voltage, unless they have m/z values that place them very near the precursor ion or on resonance with the excitation voltage in off resonance excitation experiments such as SORI. These products are observed when the activation is shifted below the m/z of the precursor but are dissociated or ejected from the trap when the activation is shifted above the m/z of the precursor. In the quadrupole ion trap, Goeringer etal. have shown that the helium bath gas actively cools product ions at a fairly high rate so that further dissociation is minimized in the quadrupole ion trap. Vachet and Glish and Doroshenko and Cotter have reported an increase in sequential dissociation when heavier bath gases such as argon, krypton, or xenon are added to the helium during CID. The general lack of sequential dissociation observed in ion trap CID may or may not be regarded as a positive result. In terms of peptide and protein sequencing, the absence of internal fragments from the amino acid chain makes deduction of the sequence from the MS/MS dissociation easier. On the other hand, some use has been made of the immonium ions of individual amino acids that appear from sequential dissociation of the precursor to help determine the amino acid content of the peptide under study.

## CONCLUSIONS

This chapter has focused exclusively on the effects of experimental variables, described in terms of the set of figures of merit, on CID. At least an equal amount could be said about the effects of ion structure on CID behaviour, even if the discussion were limited to what is known about peptides and proteins. An exhaustive summary of what is known about biological ion structural effects is beyond the scope of this chapter. The interested reader is referred to the literature on the effects of primary structure, secondary structure and conformation, and charge state on the dissociation reactions of peptides and proteins. The reader should also note that MS and tandem MS is seeing increasing use in the study of other important biological molecules such as nucleic acids, lipids, and carbohydrates.

CID is a widely applicable technique that can be used to obtain sequence and structural information from biologically derived ions. A wide variety of activation conditions lead to complementary information from different CID methods. CID conducted on sector or TOF/TOF instruments with kiloelectron-volt collision energy and low collision numbers leads to efficient dissociation of the amide bonds in peptides to yield sequence information, while also accessing amino acid side-chain cleavages to aid in the distinction of isomeric and isobaric amino acids. Dissociation spectra recorded in this high-energy regimen are extremely reproducible. A disadvantage of operation in this regimen is that dissociation rates must be driven at $10^4$–$10^6$/s, which has proven to be very difficult for ions more than approximately 3000 Da. Lower energy CID in the electron-volt regimen is effected in collision quadrupoles, where time scales can be greater by at least three orders of magnitude than at high energy. This, coupled with the high efficiency of energy transfer for collisions at lower energy, allows efficient dissociation of large ions, but without the side-chain cleavages observed at higher energy. Ion-trapping instruments can use even longer dissociation times (up to several seconds) to access very slow (1/s or less) dissociation processes. This allows ion trap slow heating methods to efficiently dissociate biomolecules of increasing size, even up to very large proteins.

# 7

# Ionic Binding to Proteins

Noncovalent binding of small molecules to proteins is a recognition *phenomenon* deeply involved in most cellular processes. Interactions with cofactors, substrates, and metal ions, as well as with other proteins, are important both in the structural *integrity* and in the activity of many proteins or enzymes. In the *biomedical* field, numerous therapeutic strategies exploit the noncovalent recognition of drug candidates by the target enzymes.

Existence of these complexes has been commonly investigated using established techniques such as gel filtration chromatography, *ultracentrifuge*, calorimetry, scattering methods (X ray *diffraction*, circular *dichroism*), fluorescence quenching, *ultraviolet* (UV) spectroscopy, or nuclear magnetic resonance spectroscopy (NMRS). Each of these techniques presents both strengths and weaknesses. Encountered difficulties include large sample requirement, long time experiments, lack of specificity of the technique for a given complex, poor mass resolution, and absence of any *spectroscopic* property of the ligand or the protein–ligand complex.

Some of these difficulties could be alleviated with the introduction of ESI MS. In fact, it appeared quite rapidly that this ionization method could allow the preservation of some noncovalent interactions during desorption in the gas phase: Intact non-covalent complexes preexisting in solution could then be characterized by ESI-MS completely desolvated in the mass spectrometer.

First examples of noncovalent complexes characterized by ESI were reported for synthetic molecules. The term *supramolecular chemistry* was coined by chemists to name these noncovalent synthetic assemblies. Accordingly, *supramolecular mass spectrometry* emerged to designate this related novel MS in which instrumentation and analytical conditions were optimized to preserve noncovalent interactions, which was absolutely unusual in classic MS. Similar applications were soon developed in the biological area. In 1991, two groups simultaneously demonstrated that by carefully controlling several experimental parameters, specific protein--ligand noncovalent interactions could survive the ESI-MS analysis. Since these initial reports, the use of ESI-MS to characterize increasingly heavy or fragile noncovalent complexes has been widely illustrated. The growing number of publications reported on the topic reflects the potential of applying this technique to solve various biological problems.

When the formation of a noncovalent complex between a protein and any metal ion or ligand is investigated, many questions arise:

- Which molecules are interacting specifically within the complex?
- What is the stoichiometry of the complex? Is this stoichiometry unique or heterogeneous?
- How tight are the interactions?
- Are the interactions specific or artifacts (simple aggregation)?
- Where are the interactions located?

Obviously, MS will not provide a high-resolution three-dimensional (3D) image of the complex formed in solution, as other techniques might do. However, most of the aforementioned issues can be addressed by appropriate ESI-MS experiments. The objective of this chapter is to describe the general MS-based strategies that can be used to retrieve information regarding the noncovalent complex formation, its stoichiometry, and its stability.

Examples cited in the following sections were selected from work performed in our laboratory on commercially available instruments; they cover a variety of applications of supramolecular MS in biology.

**Important Experimental Parameters Buffer**

The buffers commonly used in biochemistry are very unfavorable for MS. Indeed, the presence of nonvolatile salts (even at trace level) has a large impact on the quality of the spectra and can even prevent ion detection. On the other hand, usual solvents in ESI-MS (organic and acidic media) do not preserve the protein's folding, which is necessary to keep specific noncovalent interactions. For this reason, the protein has to be dialyzed against a buffer compatible with both the complex stability and the MS. Because of their reasonably high volatility, ammonium salts appeared to be good candidates, and their use for the characterization of noncovalent complexes by MS has been widely illustrated. No or few ammonium acetate adducts are observed in the mass spectra, resulting in narrow multiply charged peaks with a good mass accuracy.

***Accessible m/z Range of the Mass Analyzer***

When proteins are analyzed under non-denaturing conditions (aqueous solution and controlled pH of 6–9), the number of effective charges is greatly decreased, in comparison to that detected for the individual species under denaturing conditions (strongly acidic and organic solutions, e.g., $H_2O/CH_3CN$: 1/1, 1% HCOOH). Ions are then often detected at high mass-to-charge (m/z) ratios.

Many commercially available ESI instruments are coupled to quadrupole mass analyzers, with a fairly limited measurable m/z range (<4000 m/z). This, in many cases, constitutes a major technical limitation to achieve successful supramolecular MS experiments. However, the development of ESI instruments coupled to time-of-flight (TOF) mass analyzers has overcome this limitation, because this was already convincingly illustrated by several groups. The coupling of continuous flow ES is achieved using an orthogonal TOF. In theory, the m/z range is unlimited, and it makes no doubt that these advents will contribute to boost application of supramolecular MS in many laboratories.

***Interface Parameters (Vc, P, T)***

Interface conditions are optimized to obtain optimum sensitivity and spectrum quality while preventing dissociation of the complex. Energy transferred to ions must then be controlled in order: (i) to provide a sufficient ion *desolvation*, (ii) to ensure a good transmission of the ion beam, and (iii) to preserve the specific noncovalent interactions. The following variables affecting the ion desolvation and transmission appear to be critical: the source temperature T, the cone voltage Vc, and the interfacial pressure P. A stability diagram can be drawn for each system.

In some cases, it might be difficult to completely remove all solvent molecules from the protein complex without affecting the complex stability. In this case, the resulting mass spectrum shows variable amount of solvent *molecules*, still attached to the protein complex leading to lower sensitivity, broader ion peaks, and less accurate molecular mass measurement.

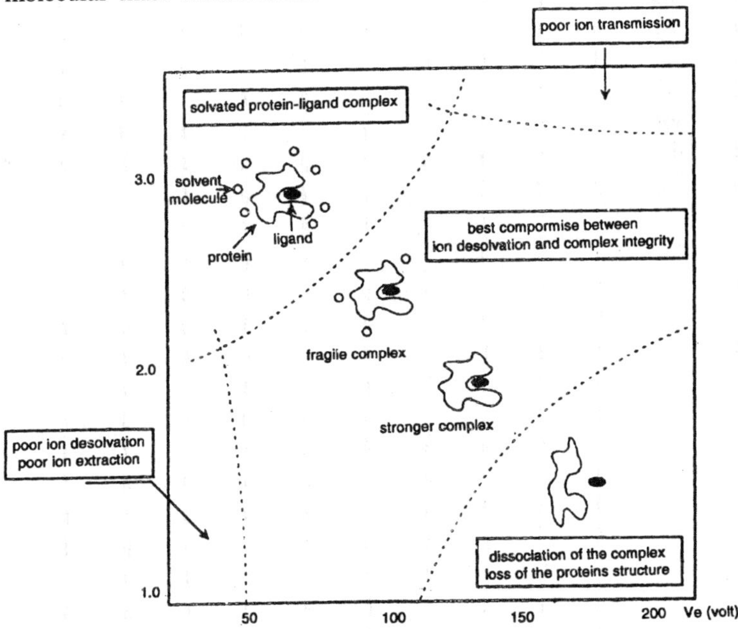


*Figure 7.1 : Schematic representation of the tuning of Vc (accelerating voltage of ions) and P (interfacial pressure) in order to both preserve the protein–ligand noncovalent interaction and detect the signal with a reasonable quality (good desolvation and transmission of ions).*

### *Introduction System*

All experiments were performed using classic *probe* at a flow rate of 4 zl/min. When a small amount of material is available, a *nanospray* probe can be used, yielding typical flow rates of 50 $\mu$l/min as shown by Robinson.

## Direct Information Provided by an ESI MS Analysis

### *Determination of the Complex Stoichiometry*

ESI MS was shown to be a rapid and sensitive tool to determine protein–ligand *stoichiometry*. Determining the number of ligands that are involved in a biologically active complex is a unique advantage of this technique. The *stoichiometry* of the complex (*n*) can be easily obtained from the direct molecular mass measurement of the complex, even if multiple stoichiometries coexist in solution:

$$\text{measured mass} = M_P + n\, M_L,$$

where $M_P$ and $M_L$ are the molecular masses of the protein and the ligand, respectively.

*Protein–Metal Interactions*

Metal ions are essential for the activity of many metalloenzymes. Counting the number of metal ions that are involved in a biologically relevant complex is an important issue and is not always straightforward with usual techniques (CD, *spectroscopic* absorption, NMRS). The potential of MS for the determination of the protein–metal stoichiometry is very promising and now well illustrated.

As an example, ESI MS has been used to investigate metal binding stoichiometry of a matrix *metalloproteinase*, the *stromelysin* 3 (ST3; molecular weight [MW] = 20,097 Da). ST3 is a zinc dependent extracellular enzyme that requires zinc and calcium for activity. Under denaturing conditions ($H_2O/CH_3CN$: 1/1, 1% HCOOH), a molecular mass of 20,097 ± 1 Da was measured for ST3, which agrees with the mass expected from the primary structure. Metal attachment can be preserved by using non denaturing conditions: In 25 mM $AcONH_4$, a unique series of multiply charged ion was observed, which could be attributed to multiple *protonated* states (8+ to 10+) of a species of mass 20,258 ± 1 Da. The mass increment of 161 Da allowed us to directly establish the *binding stoichiometry* as being stromelysin/Zn/Ca:1/2/1, which agrees with previous data obtained by other methods (accession number: P24347). This complex was shown to be resistant to gas phase collision, and mild interface conditions were not necessary for the observation of the intact complex.

Additional experiments might help if any ambiguity in the binding stoichiometry persists. For instance, intermediate stoichiometries can be revealed by coupling MS to time coarse dialyses or to the use of chelating agents (EDTA) in solution. Tandem MS experiments can also provide complementary information on the *stoichiometry*, if a gradual dissociation of the complex can be achieved. Most often, however, metal ions dissociate simultaneously in the gas phase and no intermediate products are detected.

Competitive binding to another metal ion monitored by MS can also be used to ascertain the stoichiometry of the complex. For example, comparing the mass measured under *denaturing* conditions (8889 ± 1 Da) to that measured under non denaturing conditions (9016 ± 1 Da) for the *C terminal* part of the protein p44 reveals that this protein likely binds two zinc ions (mass increment of 127 Da).

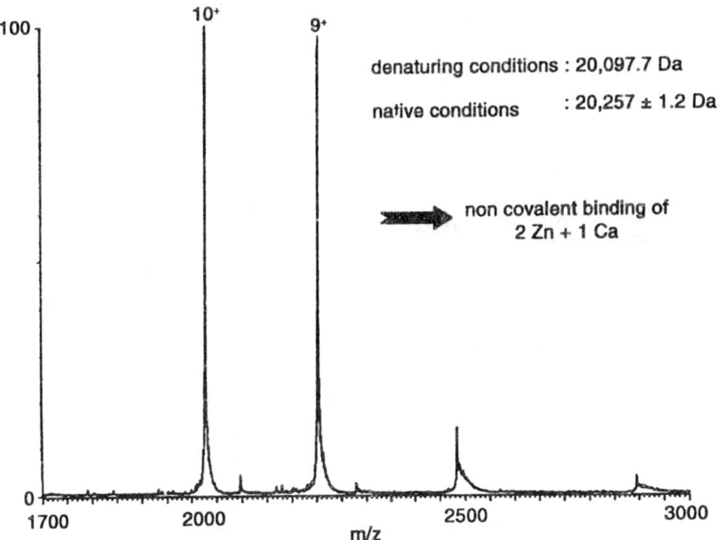


*Figure 7.2 : Determination of the metal binding stoichiometry of stromelysin 3 by electrospray ionization (ESI)-mass spectrometry (MS).*

This result is in good agreement with the zinc finger motif expected in this region of p44, particularly rich in *histidine* and *cysteine* residues, and it could also be confirmed by the successive loss of the two zinc ions observed by MS after the controlled addition of EDTA in solution (data not shown). This interpretation could be reinforced by further competitive binding experiments against several metal ions; in fact, it was found that dialysis against *cadmium ions* resulted in the replacement of the initially bound $Zn^2$ ions by two $Cd^2$ ions. This result perfectly agrees with the known similarity in the coordination behavior of these two ions.

*Protein–Ligand Interactions*

As it was previously shown for protein– metal interactions, ESI-MS can also provide information concerning the binding of substrate molecules to a protein. Comparison of the molecular masses of a species measured under denaturing and under non-denaturing conditions again indicates whether the species is or is not interacting with another molecule.

As an example, interactions between an enzyme, pig lens aldose reductase (AR, M = 35780 Da), its cofactor (NADPH, M = 742 Da), and an inhibitor (the tolrestat, M = 357 Da) have been studied by ESI-MS. AR is known to require the presence of NADPH to catalyze the reduction of its substrates (aldehydes or aldoses) to their corresponding *alcohol*. It follows an ordered addition of substrates, with NADPH binding first, and an ordered release of products, with NADP being released last. The mass spectra obtained for a preparation of AR *apoenzyme* in presence of one or both ligands. The measured molecular masses (36,512 $\pm$ 5 Da and 36,877 $\pm$ 3 Da) directly show that AR forms *binary* and *ternary* complexes, respectively, with one molecule of *cofactor* and one molecule of inhibitor.

Rigorous control experiments were performed to establish that the noncovalent complexes detected by ESI-MS were really the result of in-solution interactions rather than *artefactual* nonspecific associations occurring during the ESI process. Indeed, modifications of solution conditions, as well as complex components, would provide good support for specificity if they produce substantial change in the mass spectrum. In the case of AR, no *binary* or *ternary* complexes were observed anymore by ESI-MS after modification of the pH or after addition of organic solvent in solution, both modifications being known to destroy many specific interactions in solution. The fact that the inhibitor binds to AR only if the cofactor is already bound to its site also provides good support of specificity. *Crystal* structure confirmed this result because it clearly showed that the inhibitor site was incompletely formed in the absence of the cofactor, so that the tolrestat could not bind with a high enough affinity.

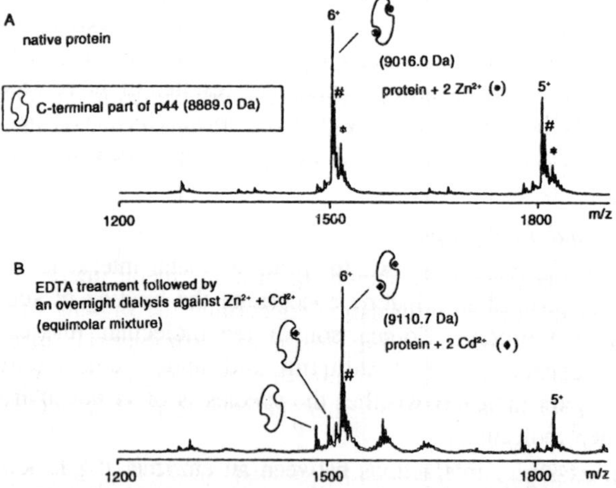


*Figure 7.3 : Metal ion content of the C terminal part of p44 (residues 321–395, M = 8889.0 Da). The p44(321–395) protein is diluted to 10 jiM in 10 mM ammonium acetate (pH 7.0).*

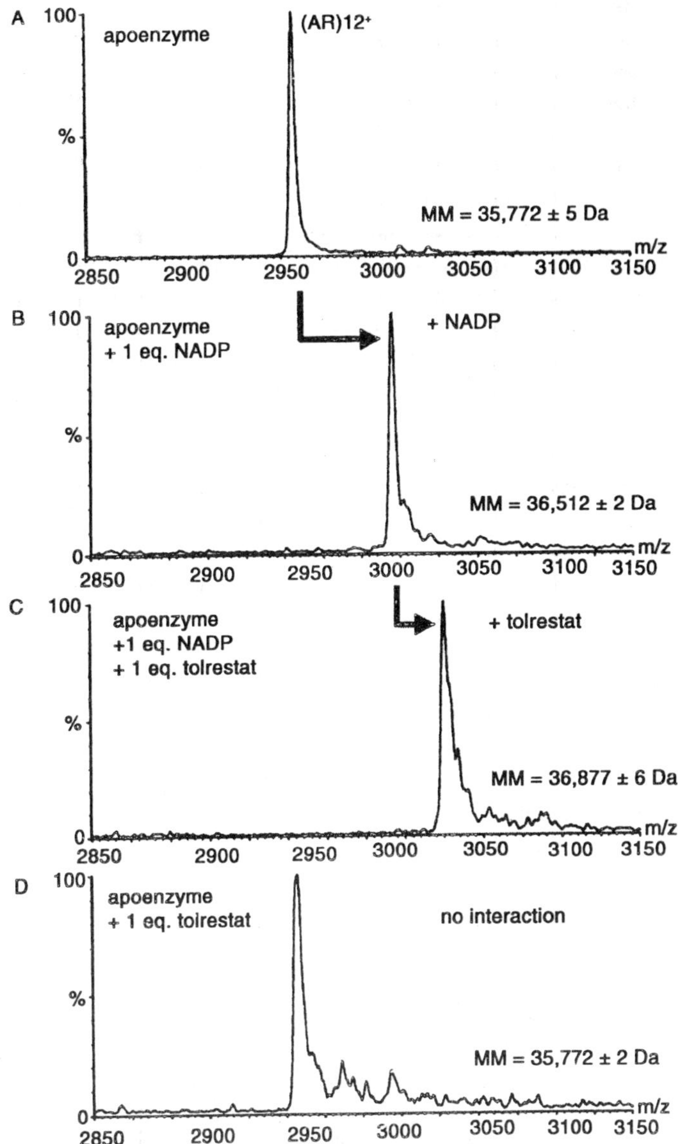


*Figure 7.4 : Electrospray ionization (ESI) mass spectra of AR in presence of (A) no ligand, (B) 1 eq of NADP, (C) 1 eq of NADP + 1 eq of tolrestat, (D) 0 eq of NADP + 1 eq of tolrestat. The binding stoichiometries are directly deduced form the measured molecular masses. The inhibitor binds to AR only when NADP is bound to its site. Spectra recorded at Vc = 50 V.*

***Determination of the Complex Stability***

Principle. *Ion intensities* (peak heights or peak area) of protein–ligand complexes observed in the ESI mass spectrum are compared to assess their binding constants in solution. Several studies have been reported that show that *quantitative* or *semiquantitative* information derived from ESI mass spectra permit the investigation of solution phase *thermodynamics* of these systems.

In semiquantitative approaches, a competitive binding experiment is performed between several *ligands* in solution, where the total protein concentration is less than the total ligand concentration. Under these competitive conditions, abundance of a protein–ligand complex depends on the value of its binding constant relative to those of other complexes. Performing these competitive binding experiments with a ligand of known binding affinity, it is possible to yield absolute values for the binding constant of the targeted complex. The use of one protein–ligand complex with a known binding constant value as internal reference allows to normalize the ES efficiency of the protein–ligand complex of interest.

A direct quantitation of all complexed and dissociated species detected on the ESI mass spectrum during a titrimetric experiment may also be used in certain cases. The ion abundance of both the bound protein and the unbound protein is then used to draw Scatchard type binding plots from which absolute binding constants can be deduced without requiring any ''*reference*'' ligand. In this case, one makes the assumption that ES efficiencies for the bound and unbound complex are similar.

*Limitations in the Determination of Binding Constants by ESI MS.*

In all the studies cited, the results obtained from ESI MS experiments were consistent with the binding constants measured by other techniques in solution. This means that for these particular systems, the mass spectra were faithfully reflecting the distribution of the species in solution.

However, this conclusion should not be thoughtlessly extended to any protein–ligand system. As pointed out by several authors, relative abundance of certain complexes may be dramatically affected by the desorption process, because of the partial loss of *hydrophobic* interactions in the gas phase. When it happens, such a distortion evidently makes the mass *spectrum* improper to investigate distribution of the species in the condensed phase.

Additional difficulties appear when one attempts to compare largely different species, because of possible differences in ionization efficiency or in ion transmission of these species. Thus, only closely related species are usually compared, for instance, complexes of a protein with a series of ligands. Note that this very often gives rise to a further *analytical* difficulty, namely the ability to separate close m/z ions.

Possible distortion arising from ion desorption or ion transfer to the mass analyzer constitutes the major risk of a false quantitative interpretation of ESI data. One then easily figures that quantitation from ESI mass spectrum should be used only with great care. General applicability of the method is far from being established, and it is crucial to investigate its validity for each new system under study.

Nonetheless, sensitivity and rapidity of ESI-MS experiments still make the technique uniquely attractive to characterize ligand binding to proteins, as this was beautifully illustrated by the group of R. D. Smith with the "BACMS" approach.

The "BACMS" Approach. The concepts of BACMS (Bio-Affinity Characterization Mass Spectrometry) approach were introduced by Smith and coworkers. BACMS exploits the high-resolution and $MS^n$ capability of Fourier transform (FT) ion *cyclotron* resonance (ICR) MS to identify tightly bound inhibitors of a target enzyme in a complex mixture. Possibilities afforded by this method were illustrated by Gao et al., with the screening of a 290-compound combinatorial library of carbonic *anhydrase* II inhibitors. Wigger et al. succeeded in identifying the preferred ligands of the HckSrc homology 2 domain protein among a library of 324 compounds in a single FT-ICR MS experiment.

A solution containing the affinity target and the ligand library prepared under competitive binding conditions is ionized by ESI. Complexes of interest are trapped and selectively accumulated in the FT-ICR. Tandem MS experiments are then performed on the isolated complexes: the resulting mass spectrum allows the dissociated ligands to be *unequivocally* identified based on their different molecular weights; moreover, relative ion intensities are measured to derive the relative binding affinities of the *ligands*. Obviously, such experiments require ionization of the ligands in the gas phase.

### *Determination of the Cooperativity in Ligand Binding to Multimeric Enzymes*

Note that in this paragraph, E designates a *multisites enzyme*, L is the ligand, and $K_n$ is the *thermodynamic* constant describing the binding of the nth ligand *molecule* to the enzyme (Eq. [1]).

$$EL_{n-1} + L \rightleftarrows EL_n \; K_n = [EL_n]/[EL_{n-1}][L] \quad (1)$$

*Definition of Cooperative Binding*

When a multimeric enzyme is composed of identical and independent (i.e., *noninteracting*) subunits, the ligand molecules are expected to be distributed *statistically* on all its binding sites. Binding constants associated to the successive binding steps (Kn) then verify Eq. (2) (s is the total number of binding sites).

$$K_{n+1}/K_n = n\,(s - n)/(n + 1)\,(s - n + 1) \quad (2)$$

When a significant deviation from the statistical binding is observed, one speaks of a cooperative binding. If the (n + 1)th ligand binds more readily than the nth ligand, the binding occurs with a positive cooperativity; the $K_n + 1/K_n$ ratio is then higher than it would be for a statistical binding. On the contrary, if the binding of the nth ligand *hinders* the subsequent binding of the (n + 1)th ligand, the binding occurs with a negative *cooperativity* and the $K_n + 1/K_n$ ratio is lower than it would be for a *statistical* binding.

***Application of Supramolecular MS to Cooperativity Studies***

*Principle*

A common approach to investigate cooperativity in binding of a ligand to a *multimeric* receptor consists of measuring the variations of the $K_n$ as the sites are progressively filled by the ligand: $K_{n+1}/K_n$ ratios are then compared to the values predicted for a statistical binding (Eq. [2]).

Because of its ability to distinguish species of different *molecular masses*, ESI MS has the potential to give a direct view of all distinct enzymatic species ($EL_{n,\, n\, =\, [0...4]}$) that are in equilibrium as increasing amount of ligand is added in solution.

A *semiquantitative* interpretation of the ESI data permits to yield relative abundances of all *enzymatic* species and to subsequently deduce the $K_n + 1/K_n$ ratios, because $K_n + 1/K_n = [EL_{n-1}] - [EL_{n+1}]/[EL_n]^2$.

***Example***

Figure elsewhere in this chapter shows illustrative ESI mass spectra recorded during addition of the oxidized cofactor $NAD^+$ to *tetrameric Sturge* on muscle *glyceraldehyde* 3 *phosphate dehydrogenase* (GPDH) or to tetiameric baker's yeast alcohol dehydrogenase (ADH).

When analyzed under non denaturing conditions in the absence of the ccfactor, both enzymes display four main charge states (z = 26 to z = 30) that match the apo form of the tetramer.

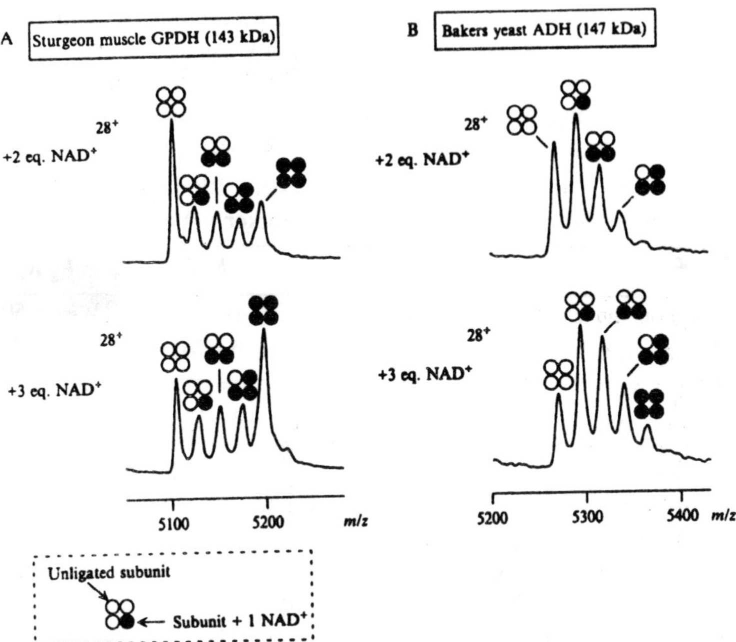


*Figure 7.5 : Typical electrospray ionization (ESI) mass spectra (here presented for the 28 fold protonated state of the tetramer) recorded during addition of NAD± to two tetrameric enzymes.*

As shown in Figure elsewhere in this chapter, addition of $NAD^+$ in solution gives rise to other species, which masses correspond to the attachment of one to four *molecules* of $NAD^+$ (M = 663.4 Da) to the *tetrameric* enzyme. These species are detected in the same charge states as the initial unbound *tetramer* (i.e., z = 26 to z = 30).

Experimental $K_{n+1}/K_n$ values deduced from the semiquantitative interpretation of the MS data for these two systems. Comparison to the values expected for a statistical binding shows that for the case of baker's yeast ADH, $NAD^+$ binding occurs without significant cooperative effect, whereas for the case of *Sturgeon muscle* GPDH, a slight positive deviation to the statistical binding is observed from site 1 to site 2 and from site 3 to site 4 (with $K_2/K_1$ and $K_4/K_3$ ratios about four times higher than that expected for a statistical binding).

***Validity of the Approach***

As mentioned previously for quantitative or semiquantitative treatment of ESI MS data, an essential prerequisite is that relative peak intensities observed in the mass spectra are reliable to investigate the solution phase *thermodynamics* (i.e., the *equilibrium* concentrations of the species). Great care in the data acquisition and in the interpretation must be taken, because it is well known that the solution-phase image might be distorted during the ESI mass analysis, especially during ion-desorption and ion-transfer processes.

**Table 7.1 : Cooperativity Studies By MS : Binding of $NAD^+$ to Sturgeon Muscle GPDH and to Baker's Yeast ADH**

| | $K_2/K_1$ | $K_3/K_2$ | $K_4/K_3$ |
|---|---|---|---|
| Sturgeon muscle GPDH | $\mu$= 1.77 | $\mu$= 0.94 | $\mu$= 1.50 |
| 38 mass spectra | $\delta$=28% | $\delta$=20% | $\delta$=28% |
| Baker's yeast ADH | $\mu$= 0.56 | $\mu$= 0.67 | $\mu$= 0.79 |
| 10 mass spectra | $\delta$=12% | $\delta$=17% | $\delta$=18% |

The following points support the validity of this approach for the cooperativity study described in the previous paragraphs:

1. The $EL_n$ species have nearly the same mass ($\Delta M \approx 664/145{,}000$) and they all have the same charge states. They are, thus, detected at very close m/z ratios, so there will not be a strong *discrimination* effect of the different species due to *focalisation* of the ion beam through the *interface* region.

2. 3D changes that may accompany the binding of NAD to the enzymatic subunits do not change *dramatically* the surface of the protein, which is exposed to the solvent; thus, $EL_n$ species likely display *comparable* solvation energies relative to the free enzyme E. Together with their close mass and charge, the assumption that their response factors to the ESI process will be similar is reasonable.

3. Experiments performed at different accelerating voltages (Vc) show that distribution of $EL_n$ species is not at all modified by *gas-phase collisions* of increasing energies. This result permits to rule out the possibility that some species may be dissociated by gas-phase collisions and be *abnormally* represented on the mass spectrum.

The major advantage of MS to study cooperativity, in comparison to established methods in the field (*UV spectroscopy*, fluorescence-quenching measurement, *capillary electrophoresis*, etc.) is to enable the view of all individual species in solution. Furthermore, rapidity and sensitivity of MS make it attractive for such studies.

*Application to the Characterization of Orphan Proteins*

An important question concerning the function of orphan proteins (i.e., proteins for which function and natural ligand are unknown) is whether their activity is mediated through ligand binding or not. Indeed, some *orphan* receptors have been shown to be constitutively active (such as CAR, ERR, and ROR), but this constitutive activity may in fact be only apparent. In this context, *supramolecular* MS seems to emerge as a new *technique* for addressing this question at the two following levels:

- To control the *homogeneity* of the expressed protein in terms of interactions with small molecules present in the *expression system* or *purification* environment.
- To identify a *potential ligand* whose activity could then be verified by an appropriate biological assay.

The ability of *supramolecular* MS to determine *existence* and *stoichiometry* of a protein–ligand complex can be used to control the homogeneity of new expressed proteins in terms of interaction with small molecules.

Indeed, some studies have shown that fortuitous ligands can be captured by the protein in the expression host with the proper stoichiometry. These molecules act as "*fillers*" and are auto-selected by the protein in the available chemical library constituted by the host cell media. They may not be the *physiological ligands*, but they prove to be essential to stabilize the active conformation of the receptors. The characterization of such ligands is then essential because if a fortuitous ligand is present, but the binding is not quantitative, the resulting chemical and conformational *heterogeneity* will hamper production and crystallization of the protein because its ligation cannot be controlled. In turn, the knowledge of the 3D structure of the ligand-binding pocket is a good starting point for the design of new high-affinity ligands. These ligands can then be used for the functional characterization of the *orphan receptor*.

If not expected and not displaying any *spectroscopic* properties, the detection of such *small ligands* bound to the protein may be very challenging. In the case of the *Ultra Spiracle Protein* (USP), the insect ortholog of the vertebrate retinoid X nuclear receptor (RXR), ESI-MS is the only technique that has been able to show the presence of an unexpected ligand molecule non-covalently attached to the receptor and to assign a molecular mass of 745 Da for this molecule. Combining ESI-MS under denaturing and native conditions, solvent extraction, and data from X-ray diffraction, it has been possible to demonstrate that the ligand captured inside the ligand-binding pocket of the USP was a *phosphatidylethanolamine*.

As the number and diversity of both clone products and host expression systems increases each day, it becomes more and more important to have in hand a technique that enables the rapid and detailed characterization of the material derived from gene expression before performing *biochemical* or *structural* studies.

## Screening of Ligands Using Mass Spectrometry

The ability of ES MS to show whether a ligand is or is not bound to a protein is now more often used in the pharmaceutical field for screening *potential drugs*. Indeed, supramolecular MS might

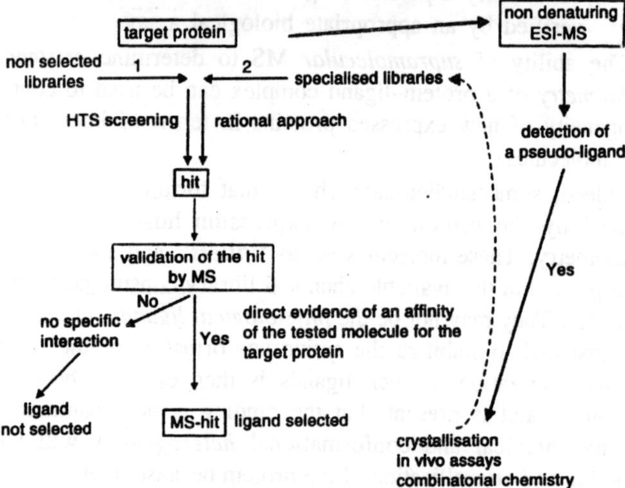


*Figure 7.6 : General strategy for the determination of the function of orphan proteins using supramolecular mass spectrometry.*

constitute a new methodology for a more reasoned and rational approach for drug discovery and for *human genome* annotation. By characterizing the interaction (positive or negative) between the target protein and a molecule (potential ligand or drug), MS will be helpful to do the following:

- Validate a hit obtained from HTS tests: an interaction between the tested molecule and the target protein will be directly observed.
- Validate a target protein evidenced by differential proteomic. Indeed, the observed variation of the protein's expression rate is not necessarily related to an affinity of the protein for the tested drug.
- Identify the physiological ligand in case of orphan proteins.

A typical screening experiment using supramolecular MS. In such experiments, library compounds are *incubated* with the target protein one by one and analyzed by ESI TOF under gentle interface conditions. Those ligands displaying an affinity for the target protein will be selected, such an affinity being evidenced by a corresponding mass shift on the mass spectrum.

In our case, the target protein is observed as a doublet because of an additional N terminal amino acid. *Incubation* conditions and ESI parameter settings are controlled using a reference ligand (ν), known to bind to the active site of the protein. After addition of 2 eq of ligand, the signal corresponding to the apo protein has completely disappeared and a new species displaying a Δm* corresponding to the mass of the reference ligand is observed. Referring to our general *strategy*, a compound will be retained as soon as a signal corresponding to a protein–ligand complex is detected. However, of major importance is the necessity to discriminate specific against nonspecific binders (both protein–ligand complexes displaying the same *molecular mass*). This can be easily performed through competition experiments between the potential ligand and the reference ligand. Two situations can occur: (i) the measured molecular mass of the resulting complex indicates that the reference ligand was able to displace the tested *compound* and (ii) the measured molecular mass of the resulting complex indicates that the reference ligand was not able to replace the tested compound. In the first case, the compound a is, thus, shown to be specific and will be definitively selected as a MS hit; in the second case, both ligands are bound to the protein, suggesting a *nonspecific binding* for the tested molecule because the protein is known to possess a single binding site. The lack of specificity of compound could have been already suspected from the multi addition effect observed.

Drug screening methods using MS have been reviewed by Siegel. *Supramolecular* MS constitutes an attractive middle throughput analytical technique that can be used as a direct detector of affinity between a target protein and a potential ligand with possible capability of *automation*.

## Information Deduced from the Gas-Phase Behaviour

During the ESI process, progressive desolvation of the droplets produces the noncovalent complexes as dried species in vacuum. On most conventional instruments, it is possible to control the *kinetic energy* of gas-phase ions, either in the first vacuum stage of the mass spectrometer, where the pressure is still of the order of 1 mbar, or in

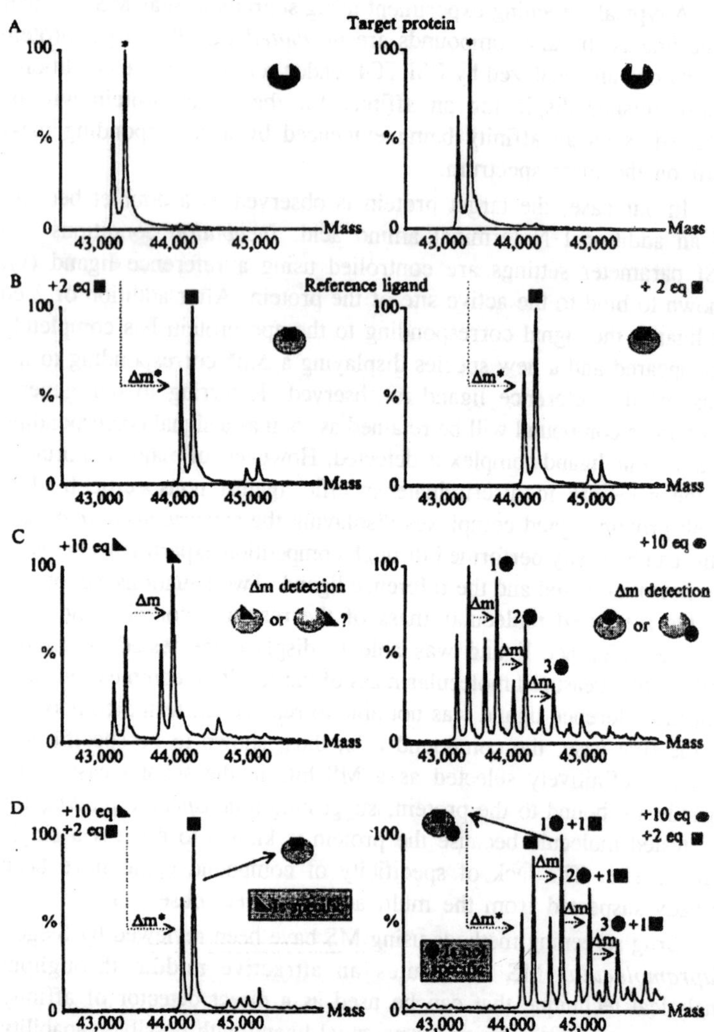


*Figure 7.7 : Typical drug screening experiment using supramolecular mass spectrometry (MS). Electrospray ionization (ESI) mass spectrum of the target protein diluted to 10μM in 50 mM annomium acetate.*

a collision-gas cell (tandem MS). Increasing the internal energy transferred to ions through *collisions* with gaseous molecules provokes disruption of the weakest interactions (i.e., the noncovalent interactions. Stability of noncovalent complexes can, thus, be measured by their resistance to *energetic* collisions as a function of either the accelerating voltage Vc or the *collision energy* (tandem MS).

**Gas-Phase Stability Measurements**

Progressive dissociation of a noncovalent protein–ligand complex can be monitored by the appearance of ions of the unbound species in the mass spectrum by increasing of the accelerating voltage Vc. An example is given in Figure elsewhere in this chapter for an AR–inhibitor complex: Going from Vc = 65 V to Vc = 85 V leads to detection of ions corresponding to the unbound enzyme (ions for the free inhibitor also appear in the low m/z range but are not shown in this figure).

A widely used criteria to evaluate gas-phase *stability* of a given non-covalent complex is the accelerating voltage needed to achieve 50% of dissociation in the mass spectrometer ($Vc_{50}$). This value is deduced from dissociation curves as those drawn on the right part of Figure elsewhere in this chapter, when abundance of the complex is equal to that of the dissociated species. Here, again, abundance of the complexed species relative to the dissociated species is measured from ion intensity displayed in the ESI mass spectrum (i.e., peak heights summed on all the charge states).

***Interpretation of Dissociation Experiments***

Relationship between Gas-Phase and Solution-Phase Stability. Several authors tried to find a correlation between the gas-phase dissociation energy (i.e., the $Vc_{50}$) and the solution-phase binding strength (i.e., the binding constants) for a series of protein–ligand *noncovalent* complexes.

For some systems, a good correlation was reported, with the order of stability in the gas phase following that in the solution phase. However, for other systems, *striking discrepancies* were found between solution- and gas-phase behavior. For a while, reasons for the observed disagreement remained unclear, even though the possible influence of hydrophobic interactions—which would be partly lost in the gas phase—was mentioned as early as 1994. However, as noted by Joseph A. Loo, "It has yet to be conclusively demonstrated that a gas-phase dissociation energy can be used to predict or even to reflect the solution-phase binding strength."

Formation of *hydrophobic* interactions in proteins is driven by water molecules surrounding nonpolar groups. These nonpolar groups tend to aggregate to minimize contacts with polar solvent molecules. This results in formation of hydrophobic interactions. Progressive

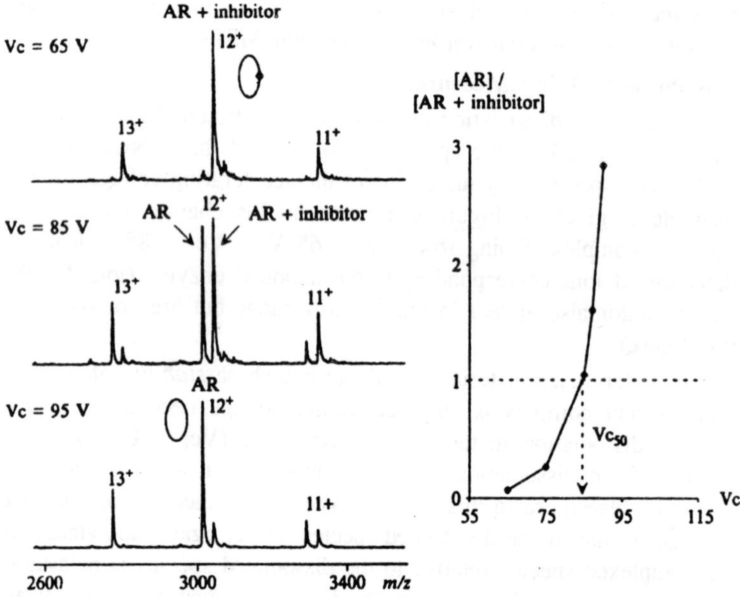


*Figure 7.8 : Gas-phase stability study of an enzyme–inhibitor complex. The enzyme is human aldose reductase (AR) complexed with its natural cofactor NADP+ (MAR-NADP = 36,880 Da); the inhibitor is a carboxylic acid of mass 345 Da. AR (10 μM in 10mM ammonium acetate) is shortly incubated with 1 eq of its cofactor and 1 eq of the inhibitor. The mixture is then continuously infused into the mass spectrometer.*

disappearance of water as species are dried in the mass spectrometer removes the driving force for hydrophobic interactions. It is, therefore, natural to assume that *hydrophobic* interactions do not contribute significantly to stability of the complex in the gas phase and do not intervene in the dissociation energy as measured by the $Vc_{50.}$ This was unequivocally verified in our group through a parallel study of several enzyme–inhibitor complexes by X-ray *crystallography* and ESI-MS.

Comparison of *Crystallographic* and Mass Spectrometric Data for a Series of Enzyme–Inhibitor Complexes. Gas-phase stability of noncovalent complexes of AR with a series of synthetic inhibitors was evaluated by ESIMS, as described earlier in this chapter. Among all inhibitors, four compounds representative of the different families of AR inhibitors were extensively studied by both ESI-MS and X-ray crystallography. The aim of this parallel study was to validate or invalidate the loss of *hydrophobic* interactions during the ion-desorption process in the mass *spectrometer*.

X-ray diffraction data allowed us to calculate the contribution of *non-hydrophobic* interactions (i.e., in the present case, sum of electrostatic and H-bond interactions) in the binding energy of each inhibitor to AR, while solution-phase measurements gave the overall binding energy (expressed by the $IC_{50}$ of each inhibitor, i.e., the concentration leading to 50% inhibition of the enzymatic activity).

The $IC_{50}$ values obtained for the four inhibitors hereby studied, the corresponding electrostatic and H-bond energies calculated from the crystallographic data ($E_{electrostatic} + E_{H\text{-}bond}$) and the $Vc_{50}$ measured by gas-phase collisions in the interfacial region of our ESI mass spectrometer. It is striking that for these four inhibitors, the $Vc_{50}$ values do not follow the order of the $IC_{50}$ values, whereas a clear correlation is observed between calculated $E_{electrostatic} + E_{H\text{-}bond}$ binding energies and the $Vc_{50}$ values.

This example demonstrates that hydrophobic contacts poorly contribute to gas-phase stability of noncovalent complexes.

**From Solution Phase to the Gas Phase: Possible Origins of Distortion**

From the preceding paragraphs, it is obvious that MS will provide an important tool for characterization of supramolecular complexes. Information as important as the stoichiometry of the complex, its stability in solution, or the nature of the interaction involved in the complex formation can be deduced from MS experiments.

**Table 7.2 : Gas-phase Stability Measurements : AR-Inhibitor Complexes**

| | $Vc_{50}$ *(volt)* | $IC_{50}$ *(nanomole)* | $E_{electrostatic} + E_{H\text{-}bond}$ *(Kcal/mol)* |
|---|---|---|---|
| AminoSNM | 70 ± 1 | 52 | –24 |
| Imirestat | 76 ± 1 | 10 | –43 |
| LIPHA3071 | 84 ± 1 | 9 | –91 |
| IDD384 | 103 ± 1 | 108 | –94 |

However, as we pointed out repeatedly in this chapter, the technique is not yet "routine" and still requires constant care. Inappropriate or uncontrolled experimental conditions, as well as the intrinsic nature of certain complexes, might be responsible for artifacts occurring in the MS analysis. If understanding of these artifacts has indeed continuously improved in the last few years, it remains essential to *systematically* perform controlled *experiments* to establish the validity of the MS results and to answer the following question: Do the data from ESI mass spectra reflect the associations existing in solution?

**Ionization Efficiency and Ion Transmission**

Species are expected to display different response factors in ESI-MS analysis depending on their charge states and solvation energies.

Solvation energy was found to play a key role in the ESI response factor of small alkaline cations. In this case, huge differences in solvation energies from one cation to the other resulted in largely unequal ion intensities detected for these ions by ESI-MS. To a lesser extent, solvation energy is also expected to play an important role in the response factor of larger systems, especially *peptides* and *proteins*. The resulting mass spectrum might then *misrepresent* the real abundance of the species in solution.

Focalization of the ion beam through the interface region may similarly discriminate some species relative to others. Significant discrimination effects between larger versus smaller m/z ratios were commonly observed by our group and others, depending on the tuning of the lenses. Comparative relative abundances between species are then not likely to be quantitative when their m/z ratios are largely different.

**Observation of Artifact Associations**

The simple observation of ions corresponding to a protein–ligand noncovalent complex does not constitute sufficient evidence of specific structurally interactions existing in solution. Control experiments are, thus, necessary to differentiate specific from nonspecific (simple aggregation) interactions that might be formed during the ESI process.

If the observed complex results from specific interactions in solution, it must be sensitive to modification of the experimental conditions affecting its stability. Modifying the conditions in solution should then produce a substantial change in the mass spectrum. For example, changing the nature of the *ligand*, *denaturing* the protein (by changing the pH, the solvent, the temperature, etc. of the solution), or diluting the sample should be reflected by the appropriate change in the intact complex abundance versus dissociated complex abundance on the ES! mass spectrum.

***Influence of the Nature of the Interactions Involved in the Complex***

From results of our group and other groups, a relationship seems to emerge between the stability of *noncovalent* complexes in the ESI process and the nature of the interactions involved in the complex formation.

In the case of leucine-zipper or of complexes between *acyl-coenzyme* A (acyl-CoA) binding protein and acyl-CoA derivatives, ES! data did not reflect the relative abundance of the noncovalent interactions in solution. The reason suggested by the authors was that interactions in these systems were mostly *hydrophobic* and were partly lost in the gas phase.

On the contrary, other complexes where electrostatic interactions were predominant displayed a unusual stability in the gas phase. For example, protein–DNA and protein–RNA noncovalent complexes appeared to be very resistant to gas-phase collisions. For certain systems like the AR-NADP system, it was shown that dissociation of the noncovalent complex could not be achieved even though the collision energy produced the breakage of a covalent bond.

Going from solution to a solvent-less environment, it is obvious that the strength of noncovalent interactions will be affected: hydrophobic effects are likely to be weakened because their driving force (water) has disappeared, but electrostatic interactions are reinforced because the dielectric constant in vacuum is 80 times lower than in an aqueous medium. Therefore, success in studying a noncovalent complex by ESI-MS will crucially depend on the nature of the interactions.

The loss of water molecules during the desorption process may thus prevent the study of certain systems by MS. This has long been considered a major drawback of this technique and even gave rise to skepticism about the future of the technique. However, it appeared that one can take advantage of this drawback and exploit it to measure the contribution of non-hydrophobic relative to hydrophobic interactions, why not use ESI-MS to assess "the type of bonding interaction that keeps complexes together?" This indeed may be of special interest in drug discovery, because specificity of ligand binding to a protein is mostly detennined by non-hydrophobic oriented interactions.

## CONCLUSION

When using carefully controlled conditions in the atmospheric pressure–vacuum interface, and when sample preparation is optimized, it is possible to preserve large specific multiprotein– metal–ligand noncovalent complexes during MS analysis. In fact, the mass measurement of complexes of more than 1 million Da is only limited by the sample preparation quality (in particular, the removal of nonvolatile salts and detergents). MS analysis can then yield precious information on the stoichiometry. In addition, in some cases, information on the relative contribution of hydrophobic and electrostatic or hydrogen bond interactions can be determined through gas phase stability experiments. It is important to keep in mind that gas-phase stability will be greater when the dominant interactions are ionic than for hydrophobically driven binding. Higher analyzer resolution will also be useful to assess more information from *mass spectra*. Obviously, ESI-MS will not provide direct structural data as NMRS or X-ray do, but the amount of material required is usually less than 1 nmol for a series of experiments.

# 8

# MASS SPECTROMETER

The world of mass *spectrometers* can be divided into two general types of instruments: ion-trapping instruments and beam instruments. The major distinguishing characteristic of ion-trapping instruments versus other types of MSs is that *tandem* mass spectrometry (MS/MS) is performed via a tandem-in-time method, rather than tandem in space. This means that each stage of mass spectrometry is performed in the same analyzer, sequentially in time. In contrast, a beam instrument, such as a triple quadrupole, has each step of MS/MS performed in different analyzers that are sequentially separated in space, one analyzer for each stage of MS/MS. In beam instruments, each subsequent stage requires the addition of another reaction region and mass analyzer. An immediately obvious advantage of trapping instruments is that multiple stages of MS/MS ($MS^n$) can be performed without instrumental modifications. The number of MS/MS stages possible in a *trapping* instrument is limited only by the ion intensity. An added benefit of trapping instruments is that 80–90% of MS/MS product ions can be trapped, whereas in linear *quadrupole instruments*, MS/MS efficiencies are typically an order of magnitude lower, and efficiencies in other beam instruments are even worse. The high MS/MS efficiency allows ion traps to perform even more stages of MS/MS. In our *laboratory*, for example, up to $MS^8$ has been performed to provide a great deal of information about the sequence of *peptides*.

There are two common types of *ion-trapping* mass analyzers: the QITMS and the ICR instrument. These two instruments operate using very different principles but have some similar characteristics. This chapter discusses the basic principles of operation of these two instruments and the variety of approaches to MS/MS that are available. It will start with the QIT, which is much less expensive and, thus, much more commonly found in *bioanalytical* laboratories. It is worth noting, however, that the ICR is a much more *mature technology* and that many of the methods used with the QIT have been adapted from ICR experiments.

## QUADRUPOLE ION TRAP MASS SPECTROMETER

### Overview

Although the QIT was patented in 1960, it was not widely used until the 1980s. Today, interest in and applications for the QIT-MS have dramatically expanded. The versatility and availability of this instrument can make the QIT-MS the workhorse of a laboratory. Typical operation of commercial ion traps gives a resolution of about 2000 and a mass range of 4000–6000 Da/charge, although the QIT-MS is capable of resolutions of more than $10^6$ and a range of 70,000 Da/charge. Whereas the figures of merit of commercial QITs are significantly lower than those for an ICR, the QITMS is operationally and mechanically much simpler. In addition to the *inherent* advantages of a trapping instrument, the QIT-MS also *boasts* relatively simple *vacuum* requirements, a small footprint, *fast analysis*, and *ruggedness*, and it is relatively inexpensive. The QIT-MS can be interfaced with a wide variety of ionisation methods, including continuous sources such as *electrospray* ionisation (ESI) and secondary ion mass spectrometry (SIMS), as well as pulsed methods like laser desorption (LD) and matrix-assisted laser desorption ionisation (MALDI). The QIT-MS also has the greatest sensitivity of any MS. These features make the QIT-MS a valuable tool for analysis of *biological* molecules.

### Trapping Ions in a QIT-MS

#### *Apparatus*

The QIT-MS operates analogously to a linear *quadrupole*, but in three dimensions rather than two. A QIT-MS is composed of three *hyperbolic* electrodes: two end-caps and one ring. In a linear quadrupole mass filter, certain combinations of *alternating* current (AC) and direct current (DC) voltages allow an ion to have a stable trajectory, which means they pass through the filter and strike a detector. In the QITMS, those ions with stable *trajectories* are trapped in the volume encompassed by the electrodes. The donut-shaped ring electrode takes the place of one pair of poles. The end-caps replace the other pair of poles. The dimensions of the QIT-MS are described in terms of the distance from the center of the trap to the closest point on each of these *electrodes*. The ring *electrode* defines the radial direction, ro. The end-cap electrodes define the axial direction, zo. One end-cap electrode will usually have a hole in it to allow ions to enter the trapping volume. In rare instances, ions may also be injected through the ring, in which case the ring electrode would have an entrance hole. Ions are typically detected by ejecting them so they will strike a standard *electron multiplier* detector. The detector is located beyond the end-cap opposite of where the ions are injected, and thus, this end-cap also has holes, to allow the ions egress.

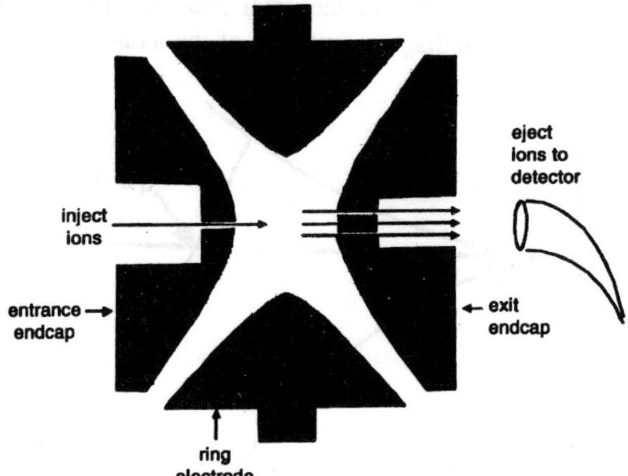


*Figure 8.1 : Cross-sectional view of the quadrupole ion trap. The entrance end-cap has one larger hole for ion injection of focused ions while the exit end-cap has several smaller holes to allow ejection.*

To generate the trapping field, an AC voltage is applied to the ring electrode, the end-cap *electrodes* or both the ring and the end-caps. In the latter arrangement, the voltage applied to the ring is 180 degrees phase shifted relative to the voltage applied to the end-cap. An optional DC voltage may be applied to either the ring or the end-caps. The movement of an ion in the trapping field is described by a second-order differential equation. A general solution to this type of differential equation was discovered by Mathieu more than a century ago. From this solution, the combinations of AC and DC voltages that result in a stable *trajectory* for an ion can be found. From Mathieu's solution, the Mathieu parameters, au and qu, are used to describe the stability of an ion in both the *axial* (z) and the *radial* (r) direction, where

$$a_z = -2a_r = \frac{-16eU}{m(r_0^2 + 2z_0^2)\Omega^2} \quad (1)$$

and

$$q_z = -2q_r = \frac{8\ eU}{m(r_0^2 + 2z_0^2)\Omega^2}. \quad (2)$$

These equations incorporate the ions' characteristics of mass (m) and charge (e), the trap dimensions both radial ($r_o$) and axial ($z_o$), the AC frequency (1), and the AC (V) and DC (U) amplitudes. The AC voltage used is in the radiofrequency (RF) range and, thus, is commonly referred to as RF. The combinations of au and qu values that result in a stable trajectory in both the axial and the radial direction are shown in the stability diagram in terms of the axial (z) direction. For

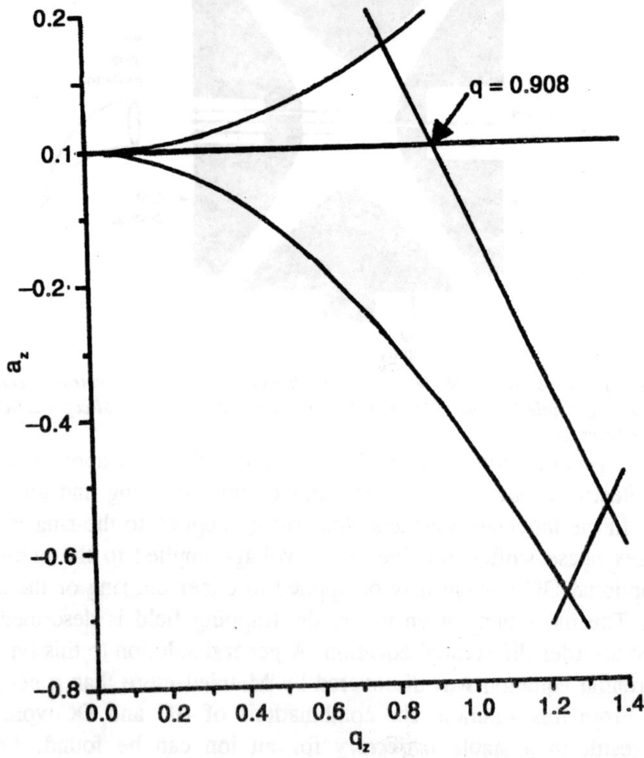


*Figure 8.2 : Stability diagram. Ions with a and q values that fall in the shaded region will have a stable trajectory and, thus, be trapped within the quadrupole ion trap mass spectrometer.*

practical reasons, ro; zo, and 1 are fixed, and ion stability is controlled by changing the *magnitudes* of the RF and DC voltages.

***Higher Order Fields***

Electric trapping fields are generated by the applied RF voltage, which creates a three-dimensional (3D) trapping field. With prescribed dimensions of $r_0$ and $z_0$ ($r^2_0 = 2z^2_0$), and a specific *hyperbolic* shape of the *electrodes*, the trapping field exerts a restoring force that increases linearly as ions move away from the center of the trap. This force pushes the ions back to the center of the ion trap when they move away and results in their trapping. However, because of field *imperfections* such as those caused by the entrance and exit holes and *intentional* distortions in zo, higher order fields are also present.

The electric field is the first derivative of the *electric* potential. Thus, the *quadrupole* potential, which is quadratic, provides a linear field, whereas a hexapole potential, which is cubic, provides a *quadratic* field. These fields affect the motion and stability of the ions in the trap. Higher order fields are weakest at the center of the trap but become more significant toward the edges (greater slope). Therefore, ion motion can most easily be described when the ions are close to

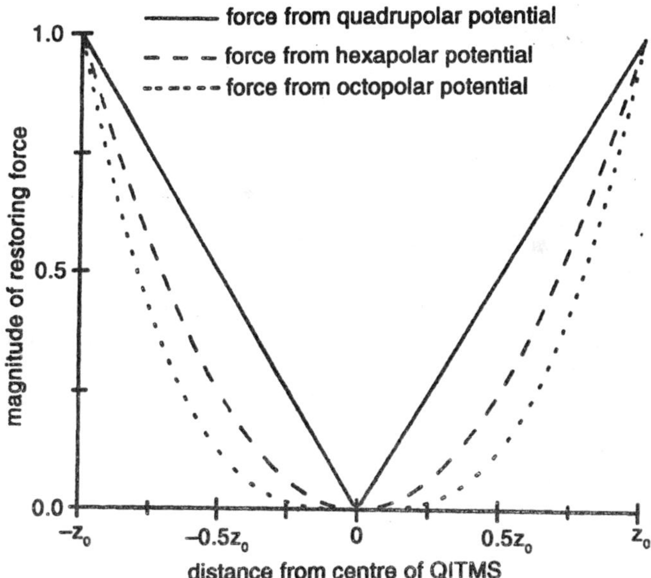


*Figure 8.3 : Magnitude of restoring forces from higher order fields as a function of distance from the center of the ion trap.*

the center of the trap, where they experience mostly linear fields.

***Resonance***

An ion is trapped in the QIT MS by maintaining stable *periodic motion* induced by the applied voltages. Each m/z ion has a unique periodicity of its motion, known as the secular frequency. The frequency in the axial direction ($w_z$) can be approximated ($q < 0.4$) in terms of the ion's a and q parameters by

$$w_z = \frac{\left(a_z^2 + \frac{q_z^2}{2}\right)^{1/2} \Omega}{2}. \qquad (3)$$

This characteristic frequency can be used to increase the kinetic energy of selected ions via resonance with a supplementary AC voltage applied across the end cap electrodes. An ion moving at the frequency of the supplementary AC voltage will gain *kinetic energy* in the axial direction.

Bath Gas. In contrast to the ICR and most other mass analyzers, the QIT MS benefits from a somewhat higher background gas pressure than is typically used for mass spectrometry. This is another advantage to the QIT MS in that pumping requirements are not as stringent, and bath gas molecules are already present for use in CID. Bath gas molecules provide collisions with the trapped ions. These low energy collisions damp the kinetic energy of the ions and restrict their movement to closer to the center of the trap. This is referred to as *collisional cooling*. At the center of the trap, the ions form a smaller cloud, and *contributions* of higher order fields are less. By condensing the ions to the center of the trapping volume, ion loss is reduced, and ions are ejected more *coherently*. Both sensitivity and resolution are improved by the presence of the bath gas. Generally, helium is the gas used, as collisional scattering is lower than with more massive gas molecules. The helium is typically added to a level of approximately 1 m Torr.

Ion-Trapping Capacity and Space Charge. The QIT-MS electrodes define a finite trapping volume. Because like charges repel each other, a finite number of ions can be trapped. It has been estimated that $10^5$ ions can be trapped at once. Whereas this capacity is more than sufficient for detection of an *analyte*, problems can arise if the trap "*fills up*" with *extraneous* ions. If the trapping limit is reached by extraneous ions, the dynamic range for analyte detection will be reduced, and in the worst case, analyte ions may not be observed at all.

The presence of a large number of ions results in an effective DC charge, known as space charge. Space charging is the result of overfilling the ion trap and begins to have an effect on performance long before the ion-trapping capacity is reached. Beyond a certain point, not only will increasing the ion accumulation time not increase the signal, but an overfilled trap also detrimentally affects the resolution of the QIT-MS. Resolution in the QIT-MS depends on ejecting all ions of the same m/z as a coherent packet to a detector. Ions in a space-charged trap will be forced by *coulombic repulsions* to spread out into a larger cloud. As the ion packet becomes more spread out, the resolution suffers. Because the m/z is determined by the RF level at the time the ions strike the detector, if the ions are spread out over time, the signal intensity of a single m/z is read over a longer time. This results not only in degradation in resolution but also as reduced signal current at the detector. The observation of broader than normal peaks is often indicative of a space-charged trap. Solutions include lowering *ionisation* times or ejecting extraneous ions, either resonantly or by increasing the RF amplitude before mass analysis.

**Ion Injection**

All discussion up to this point has involved analysis and *manipulation* of ions. However, ions must first be formed, *injected*, and successfully trapped before they can be analyzed. Previous chapters have discussed ionisation techniques. Pairing these with the QIT-MS, however, is not entirely straightforward. The complication is that these ions must have sufficient kinetic energy to enter the trap, overcoming the fringing fields created by the RF voltage that obscure the entrance. However, when ions have sufficient *kinetic energy* to enter the trap, they have sufficient energy to also exit the trap.

**Trapping Ions**

There are two possible solutions to this problem: increasing the trapping field strength or decreasing the *ions*' kinetic energies, once the ions have entered the trapping volume. Ions can be injected through a hole in either an end cap electrode (*axially*) or the ring electrode (*radially*). For methods in which the RF is increased, injection through the ring is superior. When the RF is constant, injection through an end cap is preferred.

If the RF is changed as the ions are injected, it can either be pulsed on or *ramped* to higher voltages once the ions are inside the trap. For practical reasons, *ramping* up the RF voltage is the easier technique to implement. However, only those ions in the trapping volume at the instant the RF is raised will be trapped. Therefore, these methods are ill suited for pairing with continuous ion sources such as ESI because the duty cycle is very poor. However, these methods have seen use with MALDI, a pulsed ionisation source. By timing the laser pulse with the RF, significant ion trapping can be achieved.

The more common method for trapping ions is to remove kinetic energy from ions in the trap so that the trapping fields are sufficient to hold them. Although the trapping process is not well understood, significant increases in trapping efficiencies are seen when a bath gas is present in the ion trap. In this case, ions are given sufficient *kinetic energy* to enter the trap. These ions are then thought to lose energy due to collisions with the gas molecules in the trapping volume. Sufficient kinetic energy is removed from the ions so that they are effectively trapped. This method requires a balance so that the RF level is low enough to admit the ions but high enough that the field can trap the ions once they have lost some of their kinetic energy through collisions. With a helium pressure of 1 mTorr, an RF level corresponding to a lower trapping limit of 40–50 Da/e is appropriate.

A potential problem with ion injection is that the population of ions trapped can be biased. There appears to be some *correlation* between the injected ions' m/z and the RF level of the trap during injection. By using a higher RF level during ion injection, the lighter ions are either no longer stable in the trap or not able to pass the fields at the trap entrance while the heavier ions are more effectively trapped by the stronger field. The ion population can be shifted to favour heavier ions in this way. In general, this effect is small and a single RF level during injection is generally acceptable, but caution should be used when deducing the relative solution concentrations of *species*, keeping in mind this *potential* m/z bias.

Selective Accumulation. One strength of the QIT MS is that ions can be accumulated over long periods of time to increase the number of ions present when they are ejected to the detector. However, once the trap is full, extending accumulation times will not impro ve the signal and can even be detrimental. If the analyte of interest is a small compon ent of a mixture, this can be problematic, as the trap will fill up with the *extraneous* ions. However, if the *extraneous ions* could be ejected during ion accum ulation, the trap would not fill up, and the analyte population could be increased. This can be done by resonantly ejecting the unwanted ions.

Selective accumulation can most easily be achieved using stored waveform inverse FT (SWIFT) because a large band of frequencies can be used to eject the unwanted ions. The waveform does not include those frequencies corresponding to the analytes so that these ions are not ejected. This technique can be particularly useful with MALDI, which produces a great excess of matrix ions that can quickly fill up the QIT-MS. Selective accumulation works best when the *analyte* ions are significantly different in m/z than the unwanted ions because at the RF levels used for injection, the higher mass ions have closely spaced secular frequencies. When the secular frequencies are too closely spaced, it is difficult to eject ions of unwanted m/z that are near the analyte m/z without also *ejecting analyte* ions. In such cases, ions can be injected for a period of time, the RF level raised to increase the spacing of the secular frequencies, and then the unwanted ions ejected. After ejection of the unwanted ions, another accumulation period can be used and the cycle repeated.

**Ion Detection**

Methods of Detection. In the first couple of decades after the invention of the QIT, the main mode of operation was a method termed mass selective stability. In this mode, the QIT-MS is operated *analogous* to the linear quadrupole; the parameters are set so that one m/z at a time has a stable trajectory. The trapped m/z is then ejected to the external detector. The next m/z is then trapped and ejected, and the process continues until the mass *spectrum* is acquired. This mode of operation offers little or no advantage over a linear quadrupole. Alternative modes of operation offer advantages over linear *quadrupoles*. When these modes became routine, the use of the QIT-MS was greatly increased.

***Mass Selective Instability***

The first commercial QIT-MSs were operated in a mode known as mass selective instability. In this mode of operation, the DC = 0 V ($a_z$ = 0) and an AC voltage is applied to the ring *electrode* while the end-caps are grounded, so all ions lie along the $q_z$ axis. All ions with $q_z < 0.908$ (the *boundary* of the stability diagram) are trapped. Because m/z 1 $1/q_z$ (Eq. [2]), smaller m/z ions have a larger qz and are closer to the boundary of the stability diagram. The m/z which corresponds to qz = 0.908 is the lowest mass *trapped* and is *typically* referred to as the cutoff mass. The amplitude of the RF determines this cutoff mass, so often the RF level is given in terms of the cutoff mass rather than by the voltage *amplitude*. Mass analysis can be

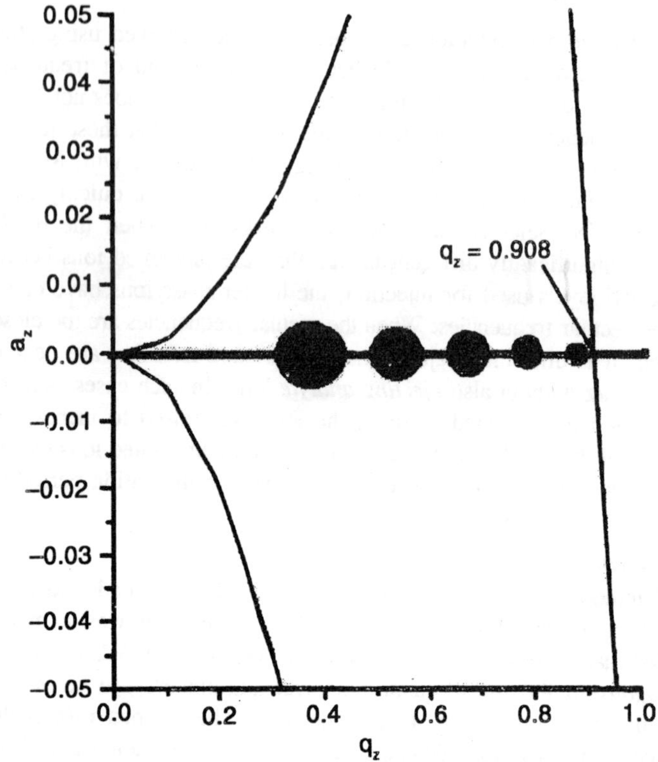


*Figure 8.4 : Mass selective instability mode. All ions lie along the qz axis with smallest ions at the highest qz values. By raising the radiofrequency level, each ion's qz value will shift up. The ions will sequentially become unstable at $q_z$ = 0.908 from smallest to largest.*

achieved by increasing the amplitude of the RF. As the amplitude increases, each ion's $q_z$ value increases. When an ion's $q_z$ value becomes just greater than 0.908, that ion's trajectory becomes unstable, and it is ejected from the trap. Low m/z ions will be ejected first, followed by higher m/z ions. An exit hole in one end cap allows the ions to escape from the trap and strike the detector. By correlating the resulting signals at the detector to the RF amplitude, the m/z of the ions can be determined. This is known as the mass selective *instability* mode of operation.

***Resonance Ejection***

The operation mode known as *resonance* ejection has made the QIT MS a practical instrument for the analysis of *biological molecules*. In the mass selective instability mode of operation, the mass range is limited to about 650 Da/e because of practical limitations in the amplitude of the RF *achievable* without electrical break down. However, the resonance ejection mode of detection can *circumvent* this problem. By applying a sufficiently large or long supplemental voltage at the resonant frequency, the ion gains enough kinetic energy to exit the trap. The first use of *resonance* ejection involved scanning the *resonance* frequency and holding the RF voltage constant. The most common implementation now involves applying a supplemental voltage at a fixed frequency and ramping the RF amplitude. Because secular frequencies are dependent upon the RF amplitude (Eqs. [2] and [3]), the secular frequencies of the trapped ions change as the RF amplitude changes. As the ions' frequencies come into resonance with the supplemental voltage, from lower m/z to higher m/z , they will be ejected. Resonance ejection can provide superior resolution and sensitivity compared to mass selective instability (as discussed in the section "*Resolution*"). More importantly, resonance ejection can be used to increase the mass range. Because of these features, resonance ejection is used on most current comm ercial QITs.

**Nonlinear Resonance Ejection**

As mentioned previously, higher order fields are naturally present in the QIT, and they are also *intentionally* induced by changes in the geometry from the *mathematically* ideal *geometry*. The higher order fields induce nonlinear resonances at specific locations in the stability diagram. One such location, due to hexapole resonance, is at a secular frequency, $w_z$, equal to one third the RF *trapping* frequency, $\Omega$. The strength of this *resonance* can be increased by increasing the hexapolar contribution. This is done by distorting the *geometry* of the trapping *electrodes* in a *nonsymmetrical* manner.

The nonlinear resonances have little effect on the ions if they are at the center of the trapping volume; however, as the ion motion moves further from the center , the *nonlinear* resonances can cause rapid ejection of the ions from the trap. Thus, by applying a supplementary voltage at a frequency of $w_z$, = $\{1/3\}\Omega$, very rapid ejection of the ions can be achieved. This *nonlinear* resonance ejection provides improved performance such as faster scan rates and better resolution. At least one commercially available QIT operates using nonlinear resonance ejection.

***Image Current Detection***

Another mode of detection is also possible in the QIT MS. Image current detection is the type of detection used in ICRs. Because each m/z has a characteristic *secular frequency*, the image current is a function of the secular *frequencies* of the *trapped* ions. The secular frequencies (and thus the m/z) in the image current can be determined via FT. This type of detection has particular advantages in that it is *nondestructive*, can detect very high m/zs, and ion populations can be remeasured to either detect a change in the ions or *achieve* better sensitivity.

Image current detection is not currently used in commercial QIT-MS instruments but has been demonstrated. A complication of this mode of detection is isolating the detector electrodes from the RF on the ring electrode. The trapping RF is large enough to obscure the ions' image currents if not properly shielded. Ions can be excited either resonantly or by a high-voltage DC pulse (100 V, 1 $\mu$s) applied to one end-cap to cause them to move *coherently*. Both of these methods function analogously to the *broadband* excitation used before detection in an FT-ICR. FT detection enables ICRs to provide vastly improved resolution and sensitivity, and this detection method could potentially enhance these performance characteristics in the QIT-MS as well. However, the best *resolutions* achieved thus far have been approximately 1000. Complications such as RF interference and space-charge-induced frequency shifts will have to be overcome before image current detection will see widespread use.

Resolution. Resolution in the QIT-MS in either mass selective instability or resonance ejection mode depends on how quickly and coherently a packet of ions can be ejected to the detector. In both of these modes, the RF amplitude is increased at a constant rate. As the ions are ejected and strike the detector, the resulting signals can be correlated to an RF level at which the ions were ejected and the mass can be determined. Resonance ejection not only increases the mass range of the QIT-MS but can also provide higher resolution *spectra*.

Scan rate is also related to resolution in resonance ejection. A faster scan rate results in lower resolution. A small amount of time is necessary for ions to become unstable and be ejected as they come into resonance with the ejection frequency. If the RF amplitude increases too quickly, all ions of a given m/z do not all have time to become destabilized equally and are not ejected together. Normal scan rates for unit mass resolution are 5000–13,000 Da/s. Slower scan rates are used to obtain higher resolution

**Ion Manipulation**

Once ions have been trapped, analysis via MS/MS can follow.

This is accomplished in three steps: isolation of the desired parent ion, activation of that ion, and detection of the product ions. Isolation of an ion is achieved using resonance ejection, which has already been discussed as a mode of ejection and detection. To isolate a pa rent ion, ions of both lower and higher m/z must be ejected. Lower m/z ions can be ejected by simply ramping the RF to a sufficient voltage so low m/z ions are ejected at $q_z = 0.908$. However, higher mass ions can be simultaneously ejected by applying a *supplemental frequency* corresponding to an m/z just higher than the parent ion's. As the RF *amplitude* increases, sequentially higher m/z ions will be resonantly ejected while those masses lower than the parention are ejected via mass selective instability. Ions can also be isolated using SWIFT in the same manner.

**Ion Activation**

Collisional Activation. Several methods can be used for ion activation, each with different advantages and limitations. The most common form of activation is by *collision* with a *neutral target* gas as part of the overall process known as collision-induced dissociation (CID). There are several methods to cause ions to undergo energetic collisions where *kinetic energy* is converted into internal energy. The excess internal energy then induces the parent ions to dissociate to product ions, from which *structural* information can be deduced.

The type of dissociation products seen with these collisional activation methods in ion traps can be different from those seen in other mass analyzers. There is a dramatic difference in the collision energies accessed in a sector instrument, typically in the 3–10 keV range, versus the 10s of electron volts in a QIT-MS. (It should be noted that the collision energy in a QIT-MS is ill-defined because the ions' kinetic energy is constantly changing due to the dynamic nature of the electric trapping field.) This difference in *magnitude* of the *collision* energy in a sector versus a QIT-MS leads to substantial differences in the amount of internal energy deposited and the internal energy distribution, with the result being notable differences in MS/MS spectra.

Although the *collision energies* in triple quadrupole instruments can be similar in magnitude to those in the QIT-MS, two other differences often lead to different MS/MS spectra being observed. First, helium is typically the collision gas in the QIT-MS, whereas argon is more commonly used in triple *quadrupoles*. With argon as the collision gas, much more *kinetic energy* can be converted to *internal energy* per collision, again leading to a different internal energy distribution. The second difference between the QIT-MS and triple quadrupole (and sector instruments) is the time frame for the reaction. Ions must dissociate (react) in 100 ms or less in the triple quadrupole instrument to be observed in the next stage of analysis.

However, in the ion trap, the times are two to three orders of magnitude longer. This longer time frame allows low energy, but *kinetically* slow reactions to occur, and can significantly favour such reactions.

***Resonance Excitation***

The first and most common method for collisional activation is resonance excitation. This method employs the same *phenomenon* of resonance as the resonance ejection method discussed previously. When a supplemental AC voltage is applied across the end cap electrodes at an ion's secular frequency, that ion's kinetic energy increases. The difference between excitation and ejection is one of degrees. By judiciously choosing the amplitude of the supplementary voltage and the time it is applied, the ion's kinetic energy can be increased without supplying sufficient energy for ejection. A typical excitation voltage would be 300–500 $mV_{p\text{-}p}$ and 10–40 ms long. Because the ions are *manipulated* by changing the amplitude of the RF, the sequence of events, or scan function, can be shown in terms of the RF amplitude.

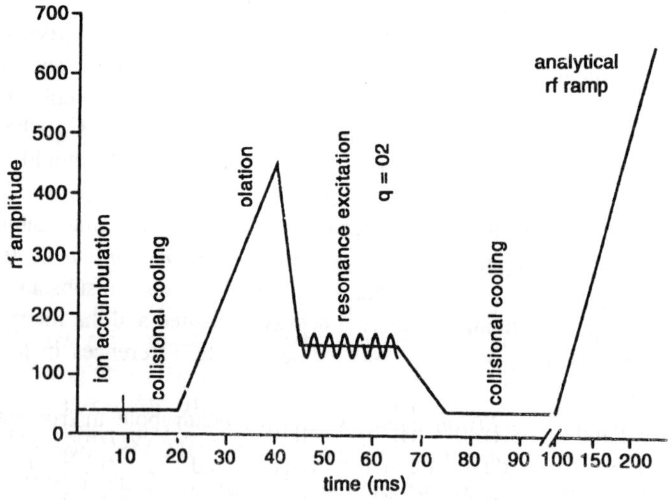


*Figure 8.5 : Scan function for MS/MS using resonance excitation collision induced dissociation.*

Unfortunately, determining the frequency needed for resonance *excitation* is not always straightforward. The equation for determining the secular frequency of an ion (Eq. [3]) is not exact. The resonant frequency can be affected by nigher order fields and can change as the ions gain *kinetic energy*. The number of *trapped ions* also affects the secular frequencies of those ions. The space charge of the ions acts as a DC potential. The effective az value no longer equals zero, and the secular frequency changes (Eq. [3]). To address this problem, some commercial QITs try to carefully control the number of ions trapped. Another approach is to use a narrow range of frequencies rather than a single frequency for resonance excitation.

Controlling the number of *trapped ions* is especially *problematic* if the ion source output *fluctuates* significantly, as can be the case with MALDI. This may be one reason that no commercial QITs offer MALDI as an ionisation option. Frequency shifts caused by the number of trapped ions are part of the motivation for finding other means of dissociation, although resonance excitation CID is still the most common method for ion dissociation.

Once the appropriate frequency is determined, the ion's kinetic energy can be increased. However, one must be cautious of the competition between ejection and excitation. If too much energy is supplied, a significant fraction of the parent ion population can be lost through ejection. The MS/MS efficiency is then decreased.

$$\text{MS/MS Efficiency} = \frac{\Sigma\ (\text{production ion intensities})}{(\text{initial parent ion intensity})}. \quad (4)$$

This not only hampers the identification of the product ions but also limits subsequent stages of MS/MS due to an insufficient number of ions.

To weight the experiment in favour of excitation over ejection, the parent ion should be trapped as strongly as possible. The ions can be thought of as being held in an energy well, called the *pseudo potential well*. The depth of this well is the maximum kinetic energy the ion can have and still remain trapped. The deeper this well, the more kinetic energy the ion can gain before it is ejected. Ions in deeper wells can, thus, undergo more energetic collisions and a greater number of collisions. This allows the ion to gain enough internal energy to dissociate before it is ejected. The depth of the well, D, is related to the $q_z$ parameter and can be approximated by the Dehmelt pseudo potential well model for $q_z$ values less than 0.4:

$$D=\frac{\left(q_z^2 m r_0^2 \Omega^2\right)}{32e} \tag{5}$$

This equation shows that ions trapped with higher qz values reside in a deeper well and can be excited to higher kinetic energies without being ejected. By increasing the RF amplitude, an ion's $q_z$ value can be increased (Eq. [2]) However, at higher qz values, lower m/z ions are not stable, so the smaller product ions are not trapped. To balance between the well depth and trapping the product ions, a $q_z$ of 0.2–0.4 is generally chosen. Thus, product ions less than about one-third the m/z of the parent ion are generally not observed.

***Heavy Gas CID***

A variation on typical resonant excitation CID involves the use of heavy gases (Ar, Xe, Kr) as the collision gas. Heavy gases offer the benefit of higher energy deposition per collision. The maximum amount of kinetic energy (Ek) that can be converted to internal energy through a collision ($E_{com}$) is given by

$$E_{com} = E_k\left(\frac{M_n}{M_n+M_p}\right), \tag{6}$$

where $M_p$ is the mass of the parent ion and Mn is the mass of the neutral molecule. For a parent ion with a given $E_k$, a heavy gas provides a higher Mn and a greater percentage of energy deposition than helium. The addition of heavy gas molecules to the QIT-MS allows higher energy deposition and a greater degree of dissociation via CID. Also, the frequency range for efficient excitation is wider with heavy gases, reducing the required precision of the secular frequency. In addition, qz values as low as 0.05 can be used, allowing low m/z products to be detected. However, the heavy gas also deteriorates the resolution and sensitivity of the QIT-MS due to greater scattering upon collision. Pulsing in the heavy gas for CID and allowing it to pump away before detection can alleviate these problems at the cost of increased time per scan (reduced duty cycle).

***Boundary-Activated Dissociation***

Methods of activation other than resonance excitation have also been explored to circumvent some of the problems. One such technique is boundary-activated dissociation (BAD). In this technique, a DC pulse is applied to the end-caps instead of an AC voltage. This causes a change in ions' $a_z$ values. As the $a_z$ value becomes large enough, the ion's a and q values approach the boundary of the stability diagram. The az value is chosen so that the ion is not ejected, but its trajectory

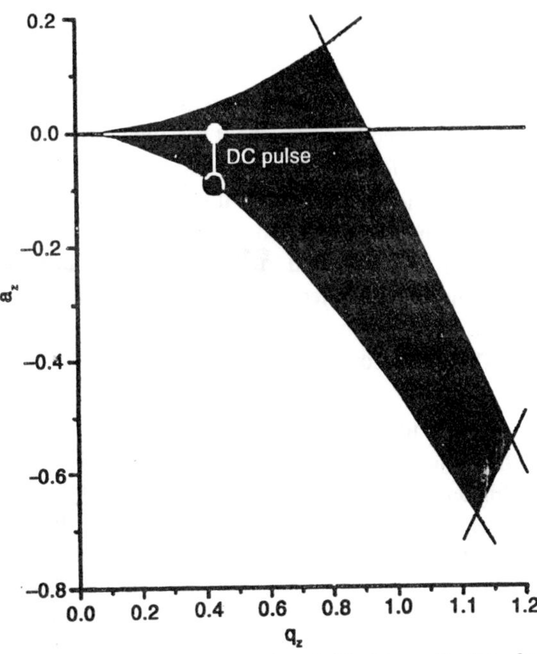


*Figure 8.6 : Boundary-activated dissociation (BAD). The application of a direct current (DC) pulse changes the ion's a value. As the a value approaches the boundary, the ion gains kinetic energy.*

increases in amplitude. As the ions' motions become larger in amplitude, there is greater force acting upon the ions, causing their kinetic energy to increase. Collisions with the bath gas molecules can then induce dissociation just as with resonance excitation. Only the magnitude of the DC must be adjusted to optimize the dissociation. As in resonance excitation, the ion must reside in a sufficiently deep well that the large-amplitude oscillations do not result in ejection. So, again, the $q_z$ values used are typically between 0.2 and 0.4, and low m/z product ions are not observed. However, the secular frequency is irrelevant in BAD, so no tuning is required, and any fluctuations in frequencies are inconsequential. Therefore, BAD is particularly useful for MS/MS when the ion intensity fluctuates significantly from scan to scan, as is common with MALDI.

*Red-shifted Off-resonance Large-amplitude Excitation*

Another activation technique that has been shown to be particularly useful for larger ions (m/z > 1000) is called red-shifted off-resonance large-amplitude excitation (*RSORLAE*). Larger ions often require a large amount of energy before dissociation is induced due to the number of degrees of freedom. Because of the competition between ejection and excitation, it can be difficult to increase the ions' kinetic energy to allow more energetic collisions without sacrificing MS/MS efficiency due to ion ejection. *RSORLAE* involves first increasing the RF *amplitude* over a period of about 10 ms, immediately dropping the RF level, and applying an excitation voltage. This voltage has a very large amplitude (21 $V_{p\text{-}p}$ compared to 400 mV in resonance excitation). However, the ions are not ejected because the frequency is shifted from the secular frequency by approximately 5% to lower frequencies (red shifted). The excitation times are variable from 30 ms to as long as 1 s.

The reason this technique can deposit larger amounts of energy in an ion without ejection lies in the higher order fields in the trapping field. These higher order fields get stronger closer to the edges of the trapping volume. The *secular frequencies* are affected by these higher order fields by being shifted to higher frequencies (*blue shifted*) as the ion moves out from the center of the trap. The initial ramping and drop in the RF amplitude serves to first compress the ions to the center of the trap under the stronger trapping field. Then, when the amplitude is *dropped*, the cloud expands, and the higher order fields have more effect on the ions. The excitation frequency is red shifted, so the ions are further excited to the edges of the trap, the blue-shifting secular frequencies move further off resonance. This allows the ions to experience a more sustained excitation period, building up internal energy, without gaining enough kinetic energy to be ejected.

RSORLAE results not only in a greater degree of dissociation than resonance excitation, but also a slightly lower $q_z$ value can be used ($q_z \approx 0.15$). Thus, somewhat smaller product ions can be trapped than those seen with resonance excitation or BAD. RSORLAE is of particular value when paired with MALDI. Because MALDI produces *predominantly* singly charged ions, ions of m/z values more than 1000 are common. Also, because the excitation frequency does not have to exactly match the *secular frequency*, *fluctuations* in the secular frequency are not as *problematic*. RSORLAE has been used to dissociate MALDI-produced ions up to m/z values 3500 with significant improvements over resonance excitation.

***Nonresonance Excitation***

One other activation technique using CID also involves a non-resonant approach. Non-resonance excitation is capable of increasing an ion's kinetic energy to 40 eV and can do so in only a few microseconds. Like RSORLAE, the ion remains trapped even with a higher kinetic energy than can be achieved via resonance *excitation*. Thus, the internal energy *deposition* can also be greater. This non-resonance excitation involves applying a low-frequency square wave (50–500 Hz) to the end-caps. Because this frequency is so low in comparison to the secular frequencies of the ions ( 100,000 Hz), this square wave can more reasonably be thought of as a series of DC pulses. The square wave causes the trapping field to change *instantaneously*. The ions will accelerate quickly to *compensate* for this change. While doing so, collisions with the bath gas convert the kinetic energy to internal energy. This process is repeated as the square wave cycles. Non-resonance excitation does not result in ejection because the kinetic energy added to the ions does not continuously increase the ions' periodic stable motions as each of the previously discussed methods does. Instead, the ions change from one periodic motion to another very quickly. The *kinetic energy* is gained as the ions adjust from one motion to the other, not as a *periodic motion* is increased in *amplitude*.

Non-resonance excitation differs from resonant excitation in that it is not selective for any one species of ions. All trapped ions will be excited simultaneously. This can be desirable or not, depending on the application. Also, because more energy can be deposited quickly by non-resonance *techniques*, higher energy dissociation channels can be accessed. Thus, non-resonance excitation can form a different group of product ions from those seen with resonance excitation.

***Surface-Induced Dissociation***

A final type of ion activation that uses collisions to increase the internal energy of ions is surface-induced dissociation (SID). However, this method differs from CID in that the collisions do not involve gas molecules. Rather, the collision is between the ion and a surface such as the ring electrode. SID has several advantages including larger energy deposition and the non-resonant nature of the excitation. Resonance *excitation* has been estimated to increase the average kinetic energy to around 9 eV, although depending on the experimental conditions, this can vary somewhat. RSORLAE and non-resonance excitation techniques can increase the kinetic energy somewhat more, to tens of electron volts. In contrast, SID kinetic energies are estimated to be in the hundreds of electron volts. The combination of such large kinetic energies and collisions with a massive surface, as opposed to light He atoms, allows SID to form products with very high threshold energy levels. The excitation is achieved via a very large DC pulse, typically 300–400 V, lasting 1–4 $\mu$s, applied to the end-caps. This pulse causes the ions to strike the ring electrode. The DC pulse must be carefully timed with the phase of the trapping RF frequency. The position of the ions upon *acceleration* must be such that the product ions are reflected into the trap and can subsequently be detected. Although this technique does show the aforementioned advantages, it also has drawbacks. Besides the experimental difficulty of implementing the fast, large, carefully timed DC pulse, SID has shown very poor efficiencies. The most abundant product ions observed are less than 5% that of the parent ion. And, as in most of the previously mentioned techniques, qz values must be raised somewhat, to 0.2–0.4 in this case, which prevents the observation of low m/z product ions.

**Infrared Multiphoton Dissociation**

Another method for inducing dissociation differs from all the previously discussed techniques in that it does not use collisions. Instead of increasing the kinetic energy of ions and then using collisions to convert that to internal energy, infrared multiphoton dissociation (IRMPD) uses photons to deposit energy into the ions. IR wavelength photons are sent into the trapping volume where they can be absorbed by the trapped ions. An IR-transparent window in the vacuum housing allows access to the trap. A laser beam can then be directed into a hole drilled through the ring electrode. An intense laser beam is used because *multiple photons* must be absorbed before an ion gains enough energy to dissociate. Either a 25- or a 50-watt $CO_2$ laser can be used, and irradiation times of approximately 100 ms are usually sufficient. IRMPD offers several advantages over other dissociation techniques. First, it is not a resonant technique, and any ion in the path of the laser will be excited. This can be used, for example, to dissociate multiple analytes simultaneously. The laser beam will also excite product ions. These ions will subsequently dissociate and can provide a richer product ion spectrum. However, the spectrum can also be more complicated and hide the genealogy of the ions. The order in which product ions are formed in this type of MS/MS is not as clear as in stepwise $MS^n$. These relationships can be revealed but require additional steps to the experiment.

One very important advantage of IRMPD is that low qz values can be used. All other excitation methods described rely on increasing the kinetic energy to increase the internal energy of ions. Therefore, the ions must be able to absorb that kinetic energy without being ejected, which requires higher $q_z$ values. IRMPD has no such restrictions, and low m/z product ions are easily trapped and observed. However, a complication with IRMPD in the QIT-MS is that collisions with the bath gas molecules between absorption events can remove energy from the ions. If collisions occur quickly enough compared to the rate of photon absorption, the ions will not reach the critical energy for dissociation. As the bath gas pressure is increased from 1 to 4 $\times$ $10^{-5}$ Torr, dissociation rates decrease. Almost no dissociation is seen at bath gas pressures above 8 $\times$ $10^{-5}$ Torr, even with a 50-Watt laser pulse lasting hundreds of milliseconds. Heating the ion trap/bath gas to 160° allows typical bath gas pressures to be used. This technique, termed thermally assisted IRMPD, can improve sensitivity by more than an order of magnitude versus IRMPD at *ambient temperature* and reduced bath gas pressure.

Gas-Phase Reactions. Another type of MS/MS reaction for which trapping instruments are particularly useful involves gas-phase ion–molecule and ion–ion reactions. The QIT-MS is particularly versatile for these types of analyses because of the combination of controlled timing of the reactions, mass selecting capabilities for both reactants and products, and $MS^n$ for both the formation of the reactants and the examination of the products. Often, these reactions are between the trapped ion and a *volatile* neutral introduced to the vacuum chamber. However, reactions between two ions have also been reported. Ion traps allow the reaction to be controlled by holding the reactants together for a variable time before ejection and detection. Therefore, studying the kinetics is straightforward, and the extent of the reaction can be well controlled.

**Ion-Molecular Reactions**

One very useful type of ion–molecule reaction for biological molecules involves hydrogen deuterium exchange. A protonated analyte reacts with a deuterated species such as $CH_3OD$, $D_2O$, or $ND_3$. As active hydrogens are exchanged for deuteriums, the m/z of the analyte is seen to increase because deuteriums are more massive than hydrogens. The number of exchangeable hydrogens can then be determined. This is of particular interest in evaluating the conformation of the analyte. For example, if a protein or peptide is folded in the gas phase, some hydrogens will be protected from exchange within the interior of the protein. However, if the same species is unfolded, more hydrogens are exposed, and the degree of exchange will be greater. H/D exchange can also be used to determine when exchangeable hydrogens are obscured by an adduct. This structural information is then used to deduce the analyte's nature in solution and how the structure is affected upon transfer to the gas phase.

*Ion-molecule* reactions that are speciûc for a certain functional group or structure can also be used to provide information about an analyte. For example, multiply charged peptides can be reacted with HI. The basic groups of the N terminus, Arg, His, and Lys, will each gain an HI adduct. By monitoring the change in m/z, the number of HI adducts can be determined and the number of basic sites deduced.

Deprotonation reactions have been used with multiply protonated protein ions. These ions can be held in an ion trap in the presence of a basic reagent such as dimethylamine. The dimethylamine molecules will extract protons from the protein, reducing its charge. The rate of deprotonation changes as the charge state changes. The change in rate reflects the different structures of the multiply charged proteins. Further, deprotonation reactions between 1,6 *diaminohexane* and CID product ions have been used to determine the charge states of these ions by observing the m/z shift as z is changed.

**Ion-Ion Reaction**

Reactions between two oppositely charged species can also be used. A QIT MS is particularly useful for these types of reactions, as both positively and negatively charged ions can be trapped simultaneously under normal operating conditions (this is not true for ICR instruments). Ion-ion reactions provide greater control in that ions are injected as needed, while neutral reactant molecules are usually present at a constant pressure. As with the ion-molecule reactions, deprotonation can also be achieved via ion-ion reactions. Ion-ion reactions have an advantage over *deprotonating ion-molecule* reactions in that anions are much stronger bases than neutral compounds. Anions can completely *deprotonate* multiply charged proteins and peptides in ion-ion reactions, whereas neutral compounds used in ion-molecule reactions typically cannot. Ion-ion *deprotonation* reactions have been useful to determine charge states and to de clutter charge state convoluted spectra. By decreasing the charge on trapped ions, the m/z values shift. Where multiple charge states of proteins overlap, shifting to a lower charge state for each protein can separate the overlapping signals. Ion-ion deprotonation of multiply charged proteins can be done using fluorocarbon anions. A similar reaction involving *protonated pyridine ions* allowed determination of the charge states of negative CID product ions. *Electron* transfer reactions have also been used to these ends.

## ION CYCLOTRON RESONANCE

The modern era of ICR has its beginning in the work of Marshall and Comisarow. By applying FT techniques, improved performance in *numerous aspects* of the experiment became possible. Today the ICR may be the most powerful mass spectrometer available. It is renowned for its *unparalleled mass* resolving power and provides *accurate mass* measurements as good as or better than any other type of mass spectrometer. The performance *capabilities* of the ICR have been steadily improving over the years as computers have become more powerful and as *stronger magnetic fields* have been obtained. Many of the performance characteristics, such as mass range and mass resolving power, are related to the strength of the magnetic field. As *magnet technology* has evolved, ICR instruments have gone from magnetic field strengths on the order of 1 Tesla, generated by electromagnets, to 3.0, 4.7, 7.0, and now 9.4 Tesla fields available with *superconducting magnets*. The desire for high magnetic fields provides two of the major contrasts between ICR and QIT MS instruments—size and cost. Whereas the QIT-MS is a benchtop instrument, an ICR requires significant laboratory floor space.

As mentioned, the performance characteristics are a function of the magnetic field strength. Current state-of-the-art instruments with 9.4-*Tesla magnets* can reach resolving powers in excess of $10^6$ for ions of m/z of 1000. The theoretical mass range exceeds $10^5$ Da/e and mass accuracies can be in the sub–part-per-million range. Like the QIT, most any type of ionisation technique can be coupled with an ICR. In many ways, the QIT-MS and ICR complement each other, and although QIT-MS instruments are found in a wide variety of laboratories, ICR is becoming an indispensable tool in state-of-the-art MS facilities involved in *biotechnology*-related areas.

### Trapping Ions in Ion Cyclotron Resonance

Principles. The main trapping force in ICR is the magnetic field. However, unlike the QIT in which the RF electric field traps ions in all three dimensions, the magnetic field traps the ions only in two dimensions, the plane perpendicular to the magnetic field. The ion motion in this plane is the *cyclotron* motion and is simply described by the basic equation of motion of a charged particle in a *magnetic field*:

$$\frac{mv^2}{r} = Bev. \tag{7}$$

in which m is the mass of the ion (kilograms), r is the radius of the ion motion (meters), v is the ion velocity (meters/s), e is the charge on the ion (coulombs), and B is the magnetic field strength (*Tesla*). Equation (7) can be rearranged to give:

$$\frac{m}{e} = \frac{Br}{v} = Bw_c, \tag{8}$$

in which $\omega_c$ is the angular velocity (*radians*/s) at which the ion is *orbiting* in the magnetic field. The angular velocity is related to *cyclotron* frequency by

$$f_c = 2\pi\omega_c \tag{9}$$

In most ICR instruments, the magnetic field is fixed, so each mass-tocharge ratio has a unique *cyclotron* frequency ($f_c$). It is worth noting that the *cyclotron* frequency of a given mass-to-charge ratio is independent of the velocity (*kinetic energy*) of the ions.

Because the magnetic field traps ions only in two dimensions, an electric field is used to trap the ions in the third dimension, the axis of the magnetic field. A potential of the same *polarity* as the charge of the ions being trapped is applied to trapping plates, which are mounted perpendicular to the *magnetic field*. Ions with axial kinetic energies less than the potential applied to the trapping plates move in a potential as well as simple *harmonic oscillators*. This trapping motion is independent of the *cyclotron motion*.

In addition to *cyclotron* and *trapping motion* of the ions in an ICR instrument, there is also a motion termed the *magnetron motion*. The magnetron motion results from the combined effect of the magnetic and electric fields. Like the cyclotron motion, the magnetron motion has a frequency associated with it. However, the *magnetron frequency* is independent of the mass-to-charge ratio of the ions. Also, whereas the cyclotron frequency is typically in the range of a few *kilohertz* to a few megahertz, the *magnetron motion* frequency is typically less than 100 Hz and is given by

$$f_m = \frac{\alpha V}{\pi a^2 B}, \tag{10}$$

where V is the magnitude of the trapping potential, **B** is the magnetic field strength, a is the distance between the trapping plates, and c is a constant related to the analyzer *cell geometry*. Thus, the magnetron motion depends not only on the magnetic and electric fields, but also on the analyzer cell geometry. The magnetron motion can be thought of as coupling with the *cyclotron motion* and displacing the center of the cyclotron motion.

***Apparatus***

Although many analyzer cell geometries have been used in ICR, the most common is the cubic cell. This cell, as the name implies, is a six-sided cube. The two sides that are orthogonal to the magnetic field are the trapping plates discussed previously. The plates on the other four sides of the cube are not important in the trapping of the ions but are used for the ion manipulation and detection processes, which are discussed below. While the cube is the simplest arrangement, a cylinder cut in quarters along the magnetic field axis is also a common geometry. Again, the main purpose of the individual sections of the cylinder is to provide the means to manipulate and detect the ions that are trapped in the cell. There have been many variations of these two basic designs, typically implemented to shim the electric field to improve various aspects of performance.

***Ion-Trapping Capacity and Space Charge***

As with QIT-MS, there is a limit to the number of charges that can be trapped in an ICR analyzer cell due to the *coulombic repulsion* between the ions. The QIT-MS can hold slightly more ions than an ICR, but the difference is not great. And just like the QIT-MS, performance is degraded in the ICR before the space charge limit is reached. Shifts in the cyclotron frequency along with peak *broadening* can occur. If ions of similar mass are present, the two peaks can coalesce into a single peak.

**Ion Injection**

Ion Sources. Since the early days of ICR, a common means to get ions into the trapping region is to form them by electron ionisation within the trapping cell. A filament to generate the *ionizing electrons* is located just outside the cell, within the *magnetic field*. This arrangement allows efficient ionisation and trapping of volatile compounds. Most *biological compounds*, however, are involatile and require a desorption ionisation technique. Desorption ionisation techniques require that the sample be external to the cell. Thus, the ions formed must be injected into the cell.

For MALDI, it is possible to locate a sample probe just outside the analyzer cell and let the ions diffuse into the cell after being formed by the laser pulse. However, the kinetic energy of the MALDI desorbed ions, which increases with mass, limits the trapping efficiency of this approach and can cause discrimination based on the ion mass.

Coupling ESI with ICR has increased challenges due to the vast

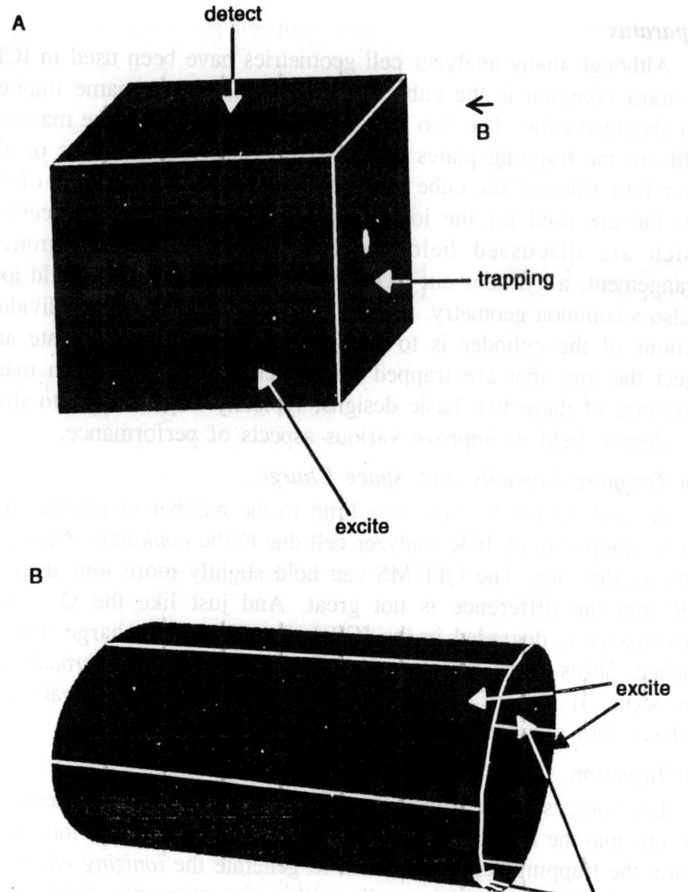


*Figure 8.7 : Schematics of (A) cubic and (B) cylindrical ion cyclotron resonance (ICR) cells.*

pressure differences between the source and the analyzer. ESI is an atmospheric pressure technique and *optimum* ICR performance is at very low pressures (e.g., $10^{-9}$ Torr or less), a pressure differential of 11–12 orders of magnitude. This can only be achieved through a number of stages of differential pumping. Although there has been one example of an ESI probe with multiple stages of differential *pumping* that can be inserted into the magnetic field near the cell, all commercial instruments have the source located outside of the magnetic field, typically at least a meter away from the cell.

### *Trapping Ions*

Injecting ions into the analyzer cell is not a trivial task. Not only do ions have to *penetrate* a large magnetic field, but they also must be slowed down enough to be trapped in the *axial dimension*. Several approaches have been used to help the ions traverse the magnetic field gradient. The earliest and probably most common approach is use of an RF only ion guide. In addition to RF only ion guides, DC wire ion guides have also been used.

After traversing the magnetic field, the next issue is trapping the ions in the cell. To get the ions into the cell, the potential on the front trapping plate (closest to the ion source) must be lower than the *kinetic energy* of the ions. To keep the ions from going straight through the cell, there must be either a higher voltage on the rear *trapping plate* (i.e., the one farthest from the ion source), or the axial kinetic energy must be decreased after the ions enter the cell. By having the axial *trapping plates* at different voltages, ions that have kinetic energies lower than the potential on the rear trapping plate will be turned around and head back in the direction of the source. If there is no change in the ion kinetic energy, the front *trapping plate* potential must be raised to a level greater than the kinetic energy of the ions. When this is done, no more ions can enter the cell. The time it takes a given ion to travel from the front plate to the maximum distance toward the rear plate (depending on the potential applied to the rear plate) and back to the front plate will be a function of the mass of the ion (smaller ions will have larger velocities). For a continuous beam of ions, this limits the number of ions that can be trapped but will not discriminate based on mass. However, for pulsed beams of ions, *mass discrimination* may occur. The gating period (i.e., the length of time that the trapping potential is lowered on the front plate to allow ions in) can be varied to optimize the trapping of a particular mass-to-charge ratio relative to the rest of the species present.

Another method for trapping ions is to decrease the kinetic energy of the ions through collisions with a background gas. This can allow longer ion accumulation times. However, the analyzer cell pressure is increased so that there is loss of resolution. A gas can be pulsed into the analyzer cell, but this substantially increases the duty cycle because of long pump-out times (20–120 s) to return to the low pressure needed for optimal resolution. As an alternative to the gated trapping or the use of collisions to trap ions, ions can be deflected off-axis in a technique known as "*sidekick*". The sidekick method provides the ability to accumulate ions in the analyzer cell for longer periods with reduced mass discrimination, increasing the number of ions that are trapped.

Whereas MALDI naturally forms a pulse of ions, ESI generates a continuous beam of ions. However, a common mode of operation of ESI is to accumulate the ions in an RF-only multipole external to the magnetic field and then inject the ions as a pulse. Because the ion injection portion of the ICR experiment is typically only a small fraction of the overall cycle time, accumulating the ions externally and injecting them in a pulse can greatly increase the effective duty cycle. One point of caution concerning external ion accumulation in an RF-only *multipole* is that it has been shown that above a certain space-charge value, dissociation of the ions can occur. This leads to a mass spectrum that is not representative of the ion species being formed in the ion source.

The ability to inject ions from external ionisation (EI) sources makes the combination of almost any ionisation technique with ICR feasible. Although MALDI and EI can readily be performed in the magnetic field near the analyzer cell, it is now common to have these ionisation sources external to the field. This allows ready access to many ionisation techniques on a given instrument. In some designs, it is possible to rather *quickly* switch from one ionisation method to another. This ability is facilitated by the fact that mass calibration is dependent only on the magnetic field, and thus, the same calibration applies regardless of the ionisation technique, in contrast to many types of *mass spectrometers*. It is also possible to form ions via EI with a filament within the *magnetic field* and, at the same time, inject ions from an external source. This allows even more *accurate mass* measurements to be obtained.

## Ion Detection

### *Image Current*

ICR is unique among MS techniques in that the ions can be readily detected without impinging them on an electron multiplier detector or any other type of detector external to the mass analysis region. They are detected inside the cell without destroying them, and therefore, after the ions have been detected, they can be manipulated further to perform more *sophisticated mass spectrometric* experiments. Two steps are involved in the ion-detection process: ion excitation and subsequent detection of the current induced on the analyzer cell plates by the ions (*image current*).

In the *basic scheme*, two of the four plates parallel with the magnetic field are used to *excite* the ions (*excitation plates*), and the other two are used to detect the ions (*detection plates*). The excitation involves applying a broadband signal to the excitation plates that spans the *cyclotron* frequency *range* of ions that are trapped. This excitation signal will cause the ions to gain kinetic energy. Because the *cyclotron* frequency remains constant, the ions will increase the radius of their orbit. By *judicious choice* of the excitation signal parameters, the orbit radius can be increased to just slightly less than the dimension of the cell. The excited ions are coherent in their motion (i.e., ions of each individual mass-to-charge move together in a packet).

If the ions are positively charged, as they approach a detection plate, *electrons* are attracted to that plate. As the packet of ions continues its orbit and approaches the opposite detection plate, electrons are attracted to that plate. A current that *oscillates* at the same frequency of the *cyclotron motion* of the ions can be detected in an external circuit between the two detection plates. This current is called the image current and is detected as a function of time. The current can then be *mathematically* processed using the Fourier equation to change from a time-domain signal to a frequency-domain signal. This process gives the technique its commonly used name, FT-ICR or FT-MS, although it should be noted that Fourier processes have been used with other types of mass spectrometers and are not unique to ICR. After the cyclotron frequency of the trapped ions has been determined, the mass-to-charge ratio can be obtained from Eqs. (8) and (9). Figure elsewhere in this chapter shows theoretical *time-domain* signals and the corresponding frequency-domain spectra.

***Resolution***

The high resolution capabilities of the ICR are based on the capability to measure the image current for a relatively long period (seconds). The longer the time domain signal can be measured, the more precisely the cyclotron frequency can be determined. The higher the precision is, the better the resolution (i.e., the separation between two peaks) unless *peak coalescence* occurs due to space charge. In addition to scaling with the acquisition time, the resolution increases linearly with increasing *magnetic field* and decreases linearly as a function of m/z.

The main limit on the resolution is the pressure in the system. The image current is damped as ions undergo collisions with neutral molecules in the vacuum system. Several effects are responsible for the decreased signal due to *ion–molecule* collisions. First, the ions'

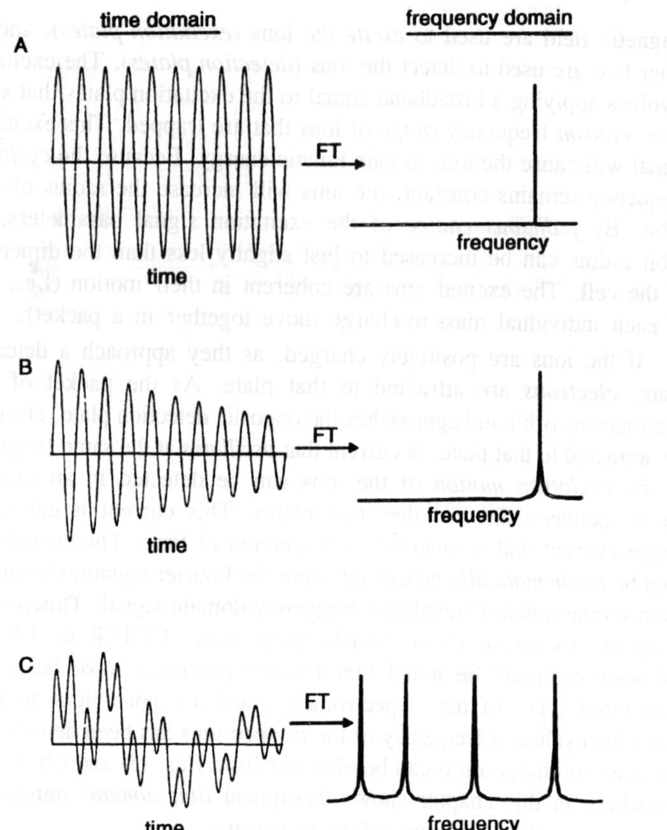


*Figure 8.8 : Examples of Fourier Transform. (A) Long time domain signal gives high resolution in frequency domain. (B) Time domain is exponentially damped (as by collisions) to give a lower frequency resolution. (C) Several frequencies convoluted in time domain are separated into frequencies by Fourier transform.*

velocity is decreased, reducing their cyclotron radius, moving them farther from the *detection plate*. Thus, there is less charge induced on the detection plate. Also, as the cyclotron radius decreases, the *magnetron radius* increases, which can cause the ions to be lost from the cell. Finally, scattering resulting from the collisions dephases the *coherent ion* motion, again reducing the charge induced on the detection plates.

The effect of pressure on resolution (*higher pressure*, *poorer resolution*) leads to one of the biggest contrasts between the operation of an ICR and a QIT-MS. In an ICR experiment, pressures in the analyzer cell are typically in the $10^{-9}$ Torr pressure range (or lower), whereas the QIT-MS is typically operated in the range of $10^{-3}$ Torr. Thus, as noted earlier, the ICR requires significantly more vacuum pumping, especially when *ionisation* techniques at *atmospheric* pressure are used.

## Ion Manipulation

### *Stored Waveform Inverse Fourier Transform*

A unique characteristic of the ion-trapping instruments is the ability to selectively manipulate the ions that are trapped. This is a result of the fact that each different mass-tocharge ratio has a unique frequency of motion in the trapping field. In the ICR, this motion is the cyclotron motion, discussed earlier. By appropriate application of an RF potential to the analyzer cell, the *cyclotron motion* can be resonantly excited, analogous to resonant excitation in a QIT-MS. The cyclotron motion is excited for ion detection but can also be excited for other purposes. A standard means for manipulating ions is known as SWIFT. In

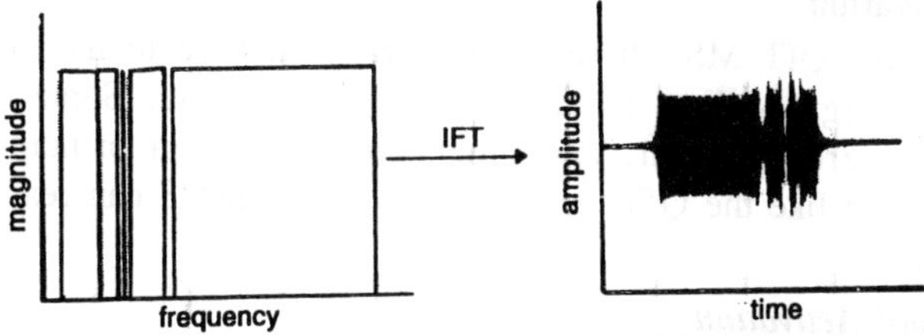


*Figure 8.9 : Stored Waveform Inverse Fourier Transform (SWIFT). The frequency domain waveform shown here will excite all ions except those with frequencies corresponding to those notched out. An inverse Fourier transform is performed to obtain a time domain signal to be applied to the excitation cell plates. The phases are modulated to stagger the phases of the frequencies to spread the power more evenly over the waveform.*

SWIFT, a frequency-domain waveform is generated to achieve the desired purpose of the ion manipulation (e.g., to eject all ions except those of a selected m/z that will be the parent ions for an MS/MS experiment). Then, an inverse FT of the frequency domain spectrum is performed to obtain a time-domain signal that can be applied to the appropriate (*excitation*) cell plates. This time-domain signal will then cause all the ions in the analyzer cell to increase in radius to the point at which they strike a plate and are lost. Very complex ion manipulation can readily be performed by this method.

### *Quadrupolar Axialisation*

Another technique used for ion manipulation is quadrupolar axialisation. This technique involves converting magnetron motion to cyclotron motion by electronic manipulation of the ions. The conversion of magnetron to *cyclotron motion* decreases the magnetron radius while increasing the *cyclotron radius*. The cyclotron radius can simultaneously be decreased by collisions, as described previously. The net effect is to focus the ions to the center of the analyzer cell. This improves the performance of an ICR instrument for a number of different parameters and experiments, such as mass resolution and accuracy, MS/MS, and ion remeasurement.

Ion remeasurement is another unique feature of ion trapping instruments. Because the ions are not destroyed when they are detected via image current measurement, they can be refocused to the center of the analyzer cell using *quadrupolar* axialisation. Once in the center of the cell, the ions can be efficiently excited and detected again. This cycle, detection followed by axialisation, can be done many times. This process can substantially improve the signal to noise ratio of a *spectrum* and the $MS^n$ efficiency.

**Ion Activation**

Like the QIT MS, there are *numerous methods* to increase the internal energy of ions to cause them to dissociate. Some of the techniques are identical, and others are similar in principle. In addition, just like the QIT, multiple stages of MS/MS can readily be performed.

***Collisional Activation***

As with all other instruments capable of performing MS/MS experiments, collisions of the selected parent ions with *neutral gas* molecules is by far the most common means of ion activation in ICR. However, a major *drawback* to this approach is the conflict between the need for a collision gas to be present to effect *collisional activation* and the need to have *ultra low pressures* to obtain the high mass resolution charac teristic of ICR ana lysis. Two approaches have been taken to address this conûict. The simplest approa ch involves pulsing the collision gas into the analyzer cell during the time period of the *collisional* activation and then delaying the detection sequence several seconds to allow the collision gas to pump out and the pressure to return to a level allowing the desired resolution to be obtained.

As an alternative to pulsing a collision gas into the *vacuum system*, a differentially *pumped dual cell* geometry can be used. In this arrangement, two cubic - trapping cells are used with a common trapping plate between them. This common plate separates regions of the vacuum system with a conductance - limiting aperture in the center of the plate. This aperture allows ions to be transferred back and forth between the two cells while keeping a pressure differential of over two orders of magnitude between the two cells. Thus, collisional activation can be performed at a higher pressure and then the produc-tions transferred to the low -pressure cell for high - *resolution mass analysis*.

The physical aspects of collisional activation in an ICR are the same as in the QIT - MS. One difference in the experiment implemen-tation is the nature of the collision gas. Typically, *helium* is used in the QIT - MS, mostly because it is already present; for other purposes, argon is more commonly used in the ICR experiment, and thus, a greater fraction of kinetic energy can be converted to internal energy in the ICR. On some occasions, this may lead to different MS/M S spectra being obtained from the two instruments.

**Gas-phase Collisional Activation**

As in the QIT - MS, the ions trapped in the ICR analyzer cell have low kinetic energies. Thus, some means to increase their kinetic energy to effect collisional activation must be used. There are a number of ways in which this can be done. The original collisional activation involved dipolar resonant excitation of the cyclotron motion, analogous to the resonant excitation that is the most common method used with the QIT-MS. The maximum *kinetic energy* that the ions can obtain during excitation is a function of the *magnetic field* strength and can be hundreds of *electron volts*. The actual kinetic energy can be calculated for this and other forms of resonant excitation.

A problem with resonant excitation in ICR is that product ions are formed with significant *cyclotron radii*, which complicates detection (and transfer between cells in a dual-cell system). There are alternative methods to standard resonant excitation that are currently used: sustained *off-resonance* irradiation (SORI), very low energy (VLE) excitation, and multiple excitation collisional activation (MECA). The most common of these methods is SORI. This is a non resonant technique in which a frequency slightly different than the ion cyclotron frequency is used to excite the ions. This difference in frequency causes the ion motion to be alternatively accelerated and then decelerated. This allows long excitation periods while keeping the ions near the center of the cell.

VLE also provides alternating periods of acceleration and decele-ration. Resonant excitation is used to accelerate the ions for a period of time, after which the *resonant* frequency is phase shifted 180°, causing the ions to be decelerated. This sequence can be repeated many times, providing a very similar result as SORI. MECA is similar to VLE except collisional cooling is used to decelerate the ions rather than doing it *electronically*.

## SURFACE INDUCED DISSOCIATION

Another activation method that avoids the use of a collision gas is SID. In this technique, the ions are forced to collide with a surface in the mass spectrometer. In the ICR, that surface is typically one of the trapping plates. A voltage pulse is applied to one trapping plate, accelerating the ions so that they strike the opposite plate. The amplitude and duration of this voltage pulse are the important parameters. Although in theory this technique is probably the easiest of all the techniques to *implement*, requiring the least amount of modification or extra equipment, it is not commonly used. This is probably because of the relatively low efficiency with which product ions are collected and because the modified surfaces, which can enhance SID, are difficult to use. However, work has been focused toward these issues, so SID may become a more widely used *technique*.

Photodissociation. Photodissociation is a logical combination with ICR. It obviates the need to add a collision gas to the instrument, and because the ions are trapped in a well defined region and at low energy, relatively good overlap between the laser beam and ions can be achieved. Both UV and IR photodissociation have been used to dissociate biological ions in ICR experiments. UV photodissociation was the first method tried for *biomolecules*. Extensive dissociation is observed, in fact so much that it is difficult to *interpret* the resulting MS/ MS spectra. As an alternative, *infrared multiphoton* photodissociation (IRMPD) has been explored. This activation method is somewhat similar to the SORI method in that the ions gain internal energy relatively slowly, so the lower energy dissociation pathways *predominate*.

***Blackbody Infrared Radiative Dissociation***

A relatively new method to dissociate ions that also relies on photons to energize the ions is termed blackbody infrared radiative dissociation (BIRD). This method basically involves heating the trapped ions by absorption of IR photons emitted (*blackbody radiation*) from the analyzer *cell plates/vacuum chamber*. By heating the vacuum system up significantly above room temperature (typically 150–200°), trapped ions can be energized to levels sufficient for dissociation to occur at observable rates. The IR *photon absorption* (heating) process of the ion is in competition with IR emission, which cools the ions.

It often takes tens to hundreds of seconds to effect dissociation by the BIRD technique. Obviously, BIRD is not a high-throughput method. However, fundamental properties of biomolecules, such as activation energies, can be determined under appropriate experimental conditions.

***Ion–Molecule Reactions***

Though sometimes not thought of as such, isolation of ions of a specific m/z and reaction of those ions with a neutral gas to form products is a classic MS/MS experiment. One of the main strengths of ICR from the outset has been the ability to use ion–molecule reactions. For many years, the main application of ICR was the study of gas phase *bimolecular chemistry*. As the instrumentation has improved and matured, ion–molecule chemistry has played a reduced role behind the high-resolution and accurate mass measurement capabilities. However, it is becoming apparent that using ion–molecule reactions for fundamental studies of biomolecules will be a great strength of ICR because of its resolution capabilities. In particular, H/D exchange of peptides and proteins to probe gas-phase conformations of these species is emerging as a major tool. Insight into the effect of parameters such as charge state can be obtained from *comparing* the number of *exchangeable protons* and the rate of exchange. A related experiment involves the stripping of protons from multiply charged peptides and proteins. Being able to control the reaction time allows kinetic measurements to be made, which provide yet another piece of data in the analysis of biomolecules.

***Electron-Capture Dissociation***

The newest activation technique that is unique to the ICR is electron-capture dissociation. In this method, low-energy electrons ($<0.2$ eV) are captured by multiply charged ions (typically proteins) in the analyzer cell. An intriguing feature of the electron-capture–induced dissociation is that the product ions formed are c and z ions that results from cleavage of the backbone amine bond, not b and y ions typically observed with other methods as a result of the *peptide* bond cleavage. In addition, *dissociation* is observed at more sites along the protein *backbone* relative to other *activation methods*.

## ACCELERATOR MASS SPECTROMETRY FOR BIOMEDICAL RESEARCH

AMS is the most sensitive method available for detecting and quantifying rare long-lived isotopes with high precision. This technique is widely employed in the earth and environmental sciences for purposes such as *radiocarbon* dating and studying the circulation of the world's oceans. It was not until the late 1980s that AMS was first used in *biological research*. Since then, it has been used primarily to investigate the *absorption*, *distribution*, *metabolism*, and excretion of radio-labeled drugs, chemicals, and nutrients, as well as in the detection of chemically modified DNA and proteins in animal models and humans. Newer applications of AMS include an isotope-labeled immunoassay and attomole-level sequencing of $^{14}C$-labeled protein, achieved by coupling Edman degradation with AMS detection. The high sensitivity of AMS measurements translates to the use of low chemical and *radioisotope* doses and relatively small sample sizes, which enables studies to be performed safely in humans, using exposures that are *environmentally* or *therapeutically* relevant while generating little *radioactive waste*.

Most biomedical AMS studies completed have employed carbon-14 as the radiolabel, although the capability exists for detecting other *isotopes* including $^{3}H$, $^{26}Al$, $^{41}Ca$, $^{10}Be$, $^{36}Cl$, $^{59}Ni$, $^{63}Ni$, and $^{129}I$. Both $^{14}C$ and $^{3}H$ are commonly used in tracing studies because they can be readily incorporated into organic molecules, either synthetically or *biosynthetically*. In this chapter, we review the principles underlying AMS-based biomedical studies, focusing on important practical considerations and experimental procedures needed for the detection and quantitation of $^{14}C$ and $^{3}H$ labeled compounds in various experiment types.

## METHODOLOGY

AMS is $10^{3}$–$10^{9}$ fold more sensitive than the decay counting methods routinely employed in *biological* studies involving *radioisotopes*. A detailed *explanation* of the differences between decay counting and AMS. Decay counting indirectly predicts the number of isotope nuclei present by measuring decay events. For $^{14}C$ and $^{3}H$ *isotopes*, this is an inefficient process, dependent on the length of time the sample is counted and the number of *nuclei* present. AMS directly quantifies each individual isotopic nucleus and is independent of these variables. This results in improved sensitivity, which means AMS biomedical studies can be performed with isotope doses 100–1000 times lower than those traditionally used and sample sizes can be $10^{5}$–$10^{6}$ fold lower.

### Instrumentation

A variety of instrument designs exist depending on the accelerating voltage and measurement requirements such as precision, resolution and the isotope range. At Lawrence Livermore National Laboratory (LLNL), a multipurpose 10 MV system is used for the analysis of a wide range of isotopes from biomedical and earth and environmental science studies. This AMS machine includes two mass spectrometers separated by an *electrostatic accelerator*, through which negative ions are accelerated to high energies. Individual *graphite* (for $^{14}C$) or titanium hydride (for $^{3}H$) pellets in *aluminum holders*, derived from the sample material, are bombarded with a large current of positive *cesium ions* (3–10 keV), which produces negatively charged elemental and *molecular ions*. These undergo an initial separation through a low energy mass spectrometer. Ions are then accelerated to the positive high voltage terminal (3–10 MV) at the midpoint of the *tandem accelerator* where they pass through a thin carbon foil or gas, which strips off *electrons*, forming positive ions and causing dissociation of interfering molecular isobars. The charge state is dependent on *ion velocity*. With a 7 MV terminal, most $C^-$ ions take on a 4+ charge and are then further accelerated back to ground potential in the second half of the accelerator. After emerging from the accelerator, the rare ions are separated from the *abundant ions* by a high energy mass spectrometer and crossed electric and *magnetic fields*. *Isotopic ions* are then individually identified through characteristic energy loss and counted in a gas ionisation detector.

AMS does not determine absolute concentrations of *radioisotopes*; each measurement is an isotope ratio: the ratio of a rare isotope to a *stable isotope* of the same element. Expressing results as a ratio corrects for fluctuations in ion transport through the AMS system. For $^{14}C$ or $3_H$ analysis, $^{13}C$ or $^{1}H$ is measured sequentially with the radioisotope of interest ($^{14}C$ or $^{3}H$, respectively) as a current in a *Faraday cup*. The isotope ratios are then compared with a set of standards with known $^{14}C$ or $^{3}H$ content to determine the precise amount of $^{14}C$ or $^{3}H$ in the sample analyzed.

At the LLNL facility, a much smaller 1 MV spectrometer was built, designed specifically for measuring $^{14}C$ in samples from biomedical studies. This lower voltage spectrometer, which can measure more than 300 samples a day with precisions of 3%, is now used for quantifying $^{14}C$ in all biological samples at LLNL. A compact AMS system dedicated to the measurement of $^{3}H$ has also been developed. The system uses a radiofrequency quadrupole (RFQ) LINAC, which has a compact size (<1.5 m) and is able to accelerate all three hydrogen isotopic species to energies sufficient for measurement using a simple magnetic spectrometer. Other compact lower voltage AMS

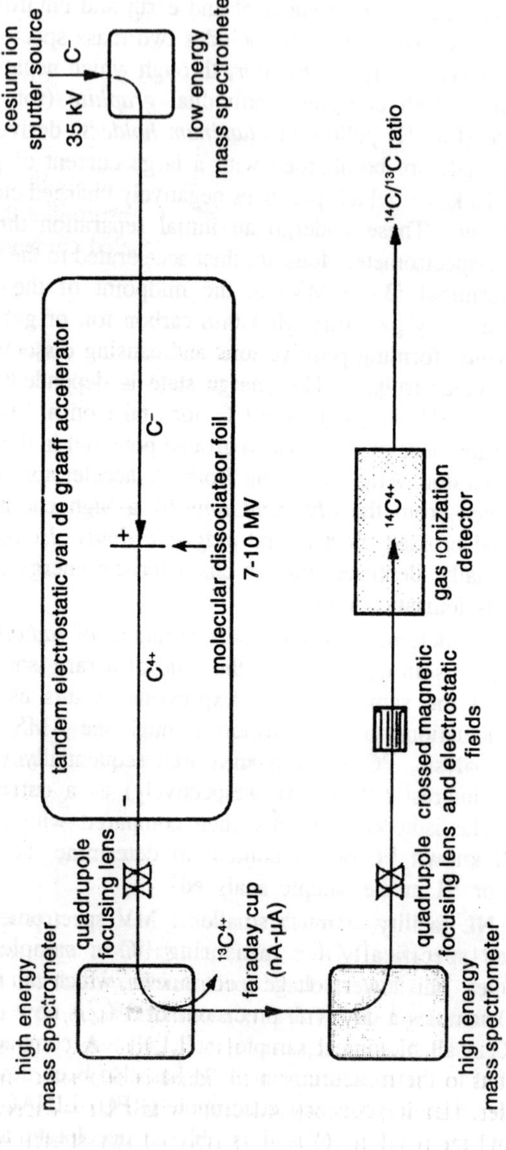


*Figure 8.10 : Diagram of the 10 MV accelerator mass spectrometry (AMS) system at the Lawrence Livermore National Laboratory.*

instruments are also being developed with variations in ion source and sample introduction.

***Sample Preparation: Methodology and Practical Considerations***

The most important considerations in preparing samples for AMS analysis are the *predicted* amount of *radioisotope* in each sample, preventing contamination and knowing the sources and amounts of any *carbon* introduced during processing. *Numerous* precautions need to be in place throughout the procedure to ensure the amount of isotope present is within the *dynamic* range of the spectrometer and to minimize the potential for contamination, to ensure that the isotope detected in the sample is associated with the labeled compound under *investigation*.

The sample of interest, for example, blood, DNA, or high performance liquid *chromatography* (HPLC) fractions, must be converted to a form that is compatible with the ion source of the instrument. Most $^{14}C$ samples are measured as *graphite* and $^{3}H$ samples as *titanium hydride*. The advantages of analyzing samples in a solid state rather than as a gas include a lower risk of cross *contamination* in the ion source, shorter memory effects when a "*hot*" sample is encountered, and higher sample throughput due to more efficient ionisation. Solid samples can be prepared at any location and sent to an AMS facility for analysis. This allows multiple researchers working on many experiments in various locations to use the instrument simultaneously. Solid samples can be left in the ion source for as long as necessary to accumulate sufficient events in the detector to attain the required measurement precision and can be saved for *reanalysis* if desired. However, the ability to directly couple an AMS instrument to an analytical HPLC, CE, or gas chromatograph (GC) system would provide a powerful and necessary technique for *biomedical* research, because it would enable online analysis of components in a sample without the need for prior separation and sample preparation. A *prototype system* has been described, which interfaces a GC to an AMS instrument. Following GC separation, *organic compounds* eluting from the column are converted to $CO_2$ in a *combustion* chamber before entering the ion source. In addition, *nonvolatile samples*, including eluent from an HPLC system, can also be analyzed using a laser induced combustion interface in which liquid is deposited into a bed of CuO powder. Heating the matrix locally with a laser causes sample combustion and the resulting $CO_2$ is transported in a stream of carrier gas into the ion source of the AMS instrument. Ultimately, the authors intend this system to be used for the online separation and detection of both $^{14}C$ and $^{3}H$ labeled compounds using various *chromatographic* techniques.

The standard procedure for the production of graphite or *titanium hydride* involves *oxidation* of the crude sample to a *gaseous mixture* including carbon dioxide and water. Carbon dioxide is then *cryogenically* extracted and reduced to form *filamentous graphite*, and water is reduced to *hydrogen gas* and adsorbed on to *titanium*, forming *titanium hydride*. The isolated sample, typically in the form of a DNA or protein solution, HPLC fraction, or wet tissue, is transferred to a quartz tube, which is baked before use to remove all bound

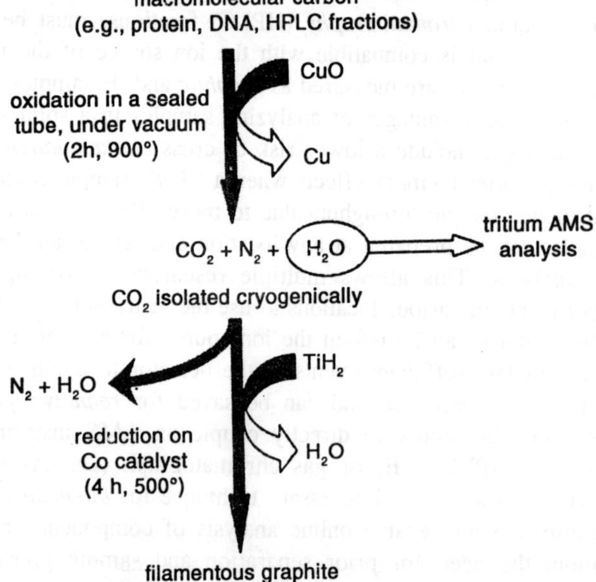


*Figure 8.11 : Stages of sample combustion and graphitisation.*

bound carbon. This tube, which must be handled with disposable forceps to prevent contamination of the exterior, is placed in an outer *borosilicate glass* tube to facilitate handling, and the volatile components of the sample are removed by *vacuum* centrifugation. *Copper oxide* in wire form is added, and the inner quartz tube transferred to a quartz *combustion tube*, which is evacuated and sealed using an *oxyacetylene torch*. The sample is then oxidized by heating in a furnace at 900° for 2 h, which produces $CO_2$, $N_2$, and $H_2O$. For production of *graphite*, the combustion tube is attached to a Y shaped disposable *plastic manifold*, the other ends of which are connected to a reaction tube, containing reagents for the reduction of $CO_2$ to elemental carbon (*titanium hydride*, *zinc*, and *cobalt powders*) and a *vacuum line*. The next step is to *cryogenically* separate $CO_2$ from the other oxidation products and trap it in the reaction tube. This involves first evacuating the reaction tube, then *clamping off* the vacuum. The combustion tube is placed in a dry *ice-isopropanol slurry*, which condenses out the $H_2O$. The end of the combustion tube is broken open, allowing transfer of $CO_2$ over to the reaction tube. This tube is placed in liquid $N_2$, which freezes out and traps the $CO_2$. Finally the hose clamp is removed to evacuate $N_2$ gas, and the reaction tube is sealed. The samples are baked in a furnace for 4 h at $500^0$. The $CO_2$ is reduced to graphite by zinc and *titanium hydride* and deposits on the *cobalt catalyst* contained in a smaller inner tube of the reaction tube.

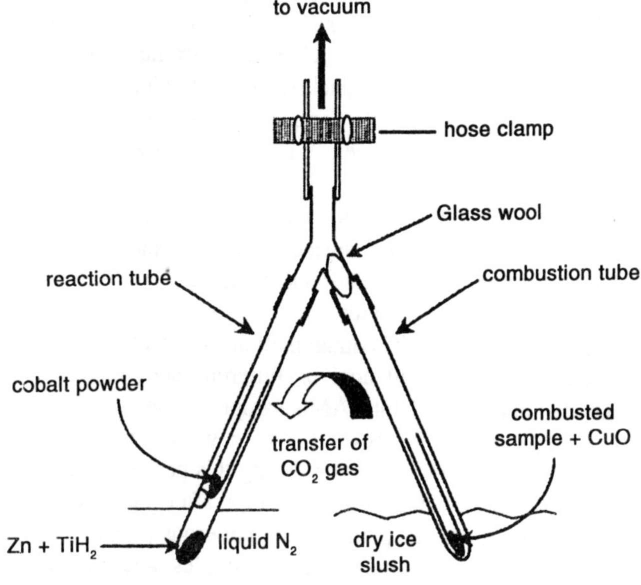


*Figure 8.12 : Transfer of CO2 from the combustion tube to the reaction tube in the preparation of graphite for* $^{14}C$ *analysis.*

The sample preparation stage is the rate limiting step in AMS studies, but a method has been developed for preparing graphite from $CO_2$ gas in septa sealed vials. This approach increases sample throughput and is less complex than the *standard protocol*, because it does not require the use of a plastic manifold or torch sealing of the transferred combustion products. Instead, the combusted sample is transferred to a septa sealed vial containing *zinc dust* and an inner quartz tube with iron powder, via a needle connected to a Luer Lok stopcock. The vial is kept in liquid nitrogen to *cryogenically trap* the $CO_2$ and $H_2O$ while *noncondensing gases* are removed by *vacuum pump*. The needle is removed from the vial, which is then placed in a block heater and held at 500°, reducing the $CO_2$ to graphite. This newer protocol results in lower process backgrounds and allows for preparation of smaller sized samples. It has been adopted as the routine method at the LLNL for preparing *biological* samples for $^{14}C$ analysis by AMS and is expected to form the basis for an integrated automated system, aimed at considerably increasing sample throughput.

For analysis of $^{3}H$, water from the combustion process is trapped using a dry ice isopropanol bath and then reduced to hydrogen gas using zinc. The resulting *hydrogen gas* is adsorbed onto titanium to produce *titanium hydride*. For AMS measurement, the prepared graphite or titanium hydride is packed into individual sample holders and placed in a sample wheel. Samples are measured at least three times, up to a maximum of seven, until the measurement precision is less than 3%.

The high sensitivity of AMS measurements means that a number of precautions must be taken during sample preparation to prevent contamination. Laboratory surfaces and equipment are *routinely* swiped using a glass filter wetted with *ethanol*, to monitor for removable contamination. Airborne $^{14}C$ contamination is detected using sorbent carbon, such as graphitized coal or fullerene soot mixed with fine metal powder and packed into AMS sample holders. These samples are left in the laboratory for several days or weeks and then measured by AMS to check for contamination during this period. These methods report airborne and surface contamination after the event, alerting users to a problem, and supplement the numerous procedures that are in place to prevent contamination of samples and the laboratory environment; each stage of the sample preparation process is *physically* segregated, according to the activity level being handled. To avoid sample contamination during handling, disposable plastic labware is used whenever possible, bench covering is changed frequently, and only specifically designated equipment (centrifuges, freezers, HPLC instruments, etc.) is used for AMS work.

**Data Manipulation**

Living entities maintain a natural level of $^{14}C$ because of the relatively constant production of this isotope in the upper atmosphere.

This level, referred to as "*Modern*," corresponds to 97.6 amol of $^{14}C$/mg of total carbon, or 6.11 fCi $^{14}C$/mg of carbon. AMS is used in biomedical studies to measure the increase in $^{14}C$ or $^{3}H$ above natural background in a particular sample, which is due to the presence of a $^{14}C$ or $^{3}H$ labeled parent compound or related derivative.

Samples that generate approximately 0.5–1.0 mg of total carbon or 2.0 mg of water upon combustion are optimal for AMS analysis using the methods described earlier. Samples containing sufficient natural carbon or $^{1}H$ are measured neat, whereas smaller samples are supplemented with a precise amount of carrier that provides sufficient carbon or water for optimal sample preparation. An ideal carrier has radioisotope concentrations well below the levels found naturally in living organisms ($<1\%$ contemporary isotope abundance). Tributyrin (25 $\mu$l of a 40 mg/ml solution in methanol or 1 $\mu$l neat tributyrin providing 0.6 mg of carbon) is often used as the carrier because it is a nonvolatile liquid and contains depleted levels of $^{14}C$ and $^{3}H$. We have found measurements of tributyrin samples to be highly reproducible. With every series of carrier supplemented samples submitted for AMS analysis, a set of three or four samples containing just the carrier is also measured.

To determine the exact amount of $^{14}C$ or $^{3}H$ in a sample by AMS measurement, the complete carbon or hydrogen inventory must be known.

For biological samples, this information can be acquired by elemental analysis or using published reference values for a particular tissue type. Calculation of $^{14}C$ or $^{3}H$ content is easiest for samples that can be obtained in quantities yielding in excess of 1 mg of carbon or 2 mg of water, such as blood, whole tissue, and isolated proteins. In this case, the excess level of 14C or $^{3}H$ in a sample, above that of an undosed control, is due to the presence of $^{14}C$ or $^{3}H$ labeled compound. If carrier is added to the samples, this must also be accounted for in the calculations by subtracting the amount of radioisotope contributed by the carrier and that contributed by the natural radioisotopic abundance of the actual sample. To do this, the precise amount of carrier added and the exact mass of biological material in the sample being analyzed must be known. Values are then converted to an appropriate form, such as picograms of radiolabeled compound per gram of tissue or tissue isolate (e.g.,. DNA or protein) using the specific activity, molecular weight of the labeled compound, and the %w/w carbon present in the tissue. For a detailed account of data handling, including example calculations.

**Applications**

AMS can be used to trace the fate of any molecule in in vitro systems or whole organisms if it is labeled with an isotope appropriate for AMS analysis. Studies have been undertaken with a variety of drugs, environmental carcinogens, and micronutrients, to determine the kinetics of absorption, distribution, and excretion, to identify and quantify metabolites, and to assess DNA or protein binding capability.

The sensitivity of AMS measurement gives this technique a number of major advantages over other methods for the detection of isotopes. Importantly, because only low doses of chemical and radioactivity are required, studies can be performed with levels of chemicals equivalent to *therapeutic* or *environmental exposures*. This is a significant feature because the biological effects observed at high doses may not extrapolate to the low doses humans typically encounter. Furthermore, with safer, low *radioisotope* doses, it is possible to perform studies in humans. Doses used in human protocols range from 10 nCi/person for metabolism and mass balance studies up to 50 $\mu$Ci/person for DNA binding studies. These are within the range that corresponds to natural background levels of radiation that individuals are exposed to during everyday life.

Although this chapter concentrates on $^{14}C$ and $^{3}H$ AMS, many other radioisotopes are increasingly being used in biological studies. Bone calcium metabolism has been studied in humans administered $^{41}C$ orally, as the carbonate, and this isotope has been investigated as a tracer for calcium uptake and deposition in cardiac ischemia. A better understanding of aluminum absorption, distribution, and clearance in healthy patients and a group with Alzheimer's disease has been gained through AMS studies using $^{26}Al$. In addition, uptake of siicic acid has been determined by $^{32}Si$-AMS. Other isotopes are also being developed for use in biomedical studies, including $^{10}Be$ and isotopes of plutonium.

**AMS Determination of Radioisotope Concentrations: Pharmacokinetic Studies**

One of the simplest and most common applications of biological AMS has been in the determination of absorption, distribution, metabolism, and excretion characteristics of isotope-labeled compounds in animal models and humans. Some unique aspects of AMS pharmacokinetic studies include the ability to determine long-term kinetics and metabolism using low doses, several months after isotope administration. Detailed pharmacokinetic data require frequent sampling, which is made possible with AMS detection, by virtue of the small sample sizes needed for analysis. AMS has been used to establish the kinetics of 3-carotene uptake and plasma clearance in a human volunteer who received a single dose of $^{14}C$-β-carotene obtained from $^{14}C$ - labeled spinach. Plasma concentrations of 3-carotene and its metabolites were determined at intervals over a 7-mo period and required just 30 $\mu$l of plasma/analysis. Such complete investigations would not be possible using other methods that lack the necessary sensitivity to detect compounds and *metabolites* months after dosing.

To quantify concentrations of $^{14}C$- or $^{3}H$-labeled compounds and their derivatives, wet tissue samples (5–10 mg) or aliquots of fluid, such as blood, plasma, or urine (typically 10–200 $\mu$l), are prepared for analysis by placing them in quartz sample tubes and drying in a vacuum centrifuge to remove water before combustion. These are then processed as usual and measured by AMS. In these types of experiments, radioisotope concentrations can be relatively high, particularly with samples taken at early time points, so it is very important to screen samples before AMS analysis to ensure the samples do not contain too much radioisotope. This is done by liquid scintillation counting (LSC) an aliquot of each sample, using an amount equivalent to that intended for AMS analysis. The AMS measurement range is 0.01–100 Modern, which equates to approximately 0.1–1350 dpm/g of carbon. Consequently, if radioisotope can be detected above background levels using LSC, the sample must be diluted before AMS analysis. If the signal is below or at the limit of detection, analysis can be performed.

A new application of AMS, made possible by advances in tritium detection, is dual isotope-labeling studies. Dingley et al. have demonstrated that two independent compounds can be traced if one is labeled with $^{14}C$ and the other $^{3}H$. For example, the liver concentrations of two heterocyclic amines $^{3}H$-PhIP and $^{14}C$-MeIQx were determined when the compounds were administered to rats individually and in combination, to ascertain whether co-administration affected bioavailability. The methodology permits liver samples from a single source to be assayed in parallel for the presence of either $^{14}C$ or $^{3}H$. Furthermore, simultaneous extraction of $CO_2$ and $H_2O$ from the same sample is also now possible so that both isotopes can be quantified using just one sample, thereby increasing sample throughput and decreasing the amount of material required. This will enable animal and human studies to be carried out involving mixtures of compounds, to more closely mimic *typical human* exposures.

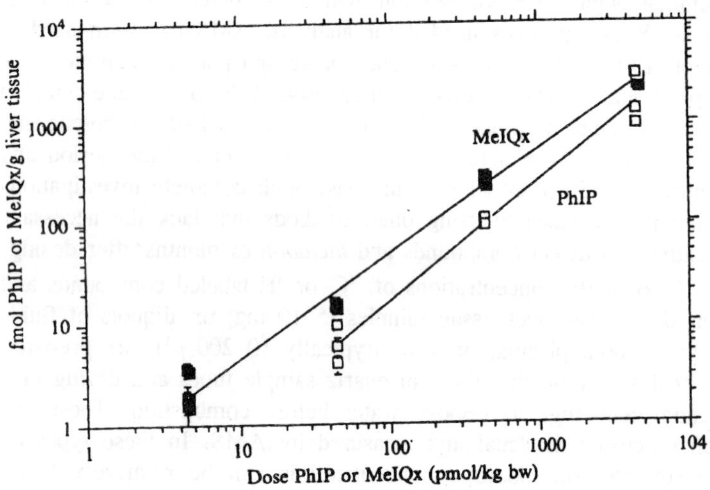


*Figure 8.13 : Concentration of $^{3}H$-PhIP and $^{14}C$-MeIQx in the liver tissue of rats administered a dose range of either $^{3}H$-PhIP or $^{14}C$-MeIQx alone or in combination.*

## HPLC AMS

AMS measurements provide no structural information, so any sample characterisation or identification must be performed before sample preparation, using chromatographic (TLC, HPLC, size exclusion, and affinity) or electrophoresis (gel and capillary) separation techniques. Coupling an HPLC step to AMS detection affords both specificity and high sensitivity, in an approach that has been used for metabolite profiling and quantification, as well as the identification of chemically modified nucleosides in humans and animal models where other techniques are not sensitive enough to detect the low levels formed. Metabolites can be extracted and analyzed from a variety of biological matrices such as plasma, urine, tissues, and milk and a full metabolite profile can be obtained using just 100 zl of urine. Before beginning an experiment, aliquots of HPLC buffers, solvents, and any standards that might be run should be measured by AMS to check for contamination. Importantly, the HPLC solvents must be compatible with the graphitisation process, in particular buffers containing sodium salts should not be used. Ideally, the buffer should not introduce an extra source of carbon, although it is sometimes possible to use such buffers if an isocratic system is employed and the amount of carbon in each fraction is constant throughout the run. In this situation, it is important that the samples do not contain too much carbon, because this can cause the tubes to explode during the combustion step. Samples are separated by HPLC and individual fractions collected at known intervals, typically every 20–60 s. In most cases, the fractions usually contain very little carbon after removal of the solvent, and tributyrin is added to each fraction to provide a fixed amount of carbon, sufficient for AMS measurement. The quartz sample tubes have a maximum capacity of approximately 300 $\mu$l, limiting the volume that can be submitted. Therefore, in situations in which the radioisotope content is expected to be low, fractions should be concentrated before this stage. Alternatively, 300 $\mu$l aliquots can be dried down repeatedly in the quartz tubes, but this can be time consuming. Performing and analyzing a control run of unlabeled material processed in an identical manner to the test sample is vital because the background $^{14}$C or $^{3}$H level of the system determines the detection limit of the method. The concentration of $^{14}$C or $^{3}$H in each fraction is determined from the isotope ratio and is used to reconstruct a chromatogram. Individual metabolites and compounds can then be identified based on retention times and comparison to authentic standards.

As has been demonstrated for folic acid, the high sensitivity of AMS analysis can reveal the formation of previously unidentified metabolites that are formed at levels below the limits of detection achievable with traditional techniques. A further feature is the capability to detect metabolites formed after low-dose exposures, using doses of labeled chemicals comparable to therapeutic or environmental levels. This methodology was applied in a study designed to assess the distribution of PhIP and its metabolites in lactating female rats and their pups after administration of a single dose of $^{14}$C-PhIP, equivalent to an average human daily dose of this compound. The detection of PhIP and PhIP metabolites in the breast milk, stomach contents, and liver tissue of the pups by AMS provided evidence that suckling pups are exposed to this carcinogen even at low doses. To further examine the types and concentration of metabolites excreted in breast milk, metabolites were extracted by adding methanol and centrifuging the milk samples to remove protein. The supernatant was evaporated to dryness, the residue dissolved in 0.1 % trifluoroacetic acid, and subject to reverse-phase HPLC separation using a trifluoroacetic acid/ acetonitrile mobile phase. This procedure achieved greater than 95% recovery of PhIP and its metabolites, based on the radiocarbon levels of samples measured before and after extraction. Fractions were collected at 1-min intervals throughout the run. These were then concentrated and redissolved in water/methanol before adding tributyrin and preparing for AMS analysis. Reconstruction of the HPLC chromatogram using the AMS data revealed the presence of three metabolite peaks in addition to the parent PhIP molecule. These were identified by comparison to metabolites present in milk extracts from a rat treated with a high (10 mg/kg) dose of PhIP, which were fully characterized by HPLC-MS. This study demonstrates the importance of being able to investigate low-dose metabolism, because although the metabolite profile over the dose range examined was qualitatively similar, two glucuronide conjugates formed at high doses (10 mg/kg) of PhIP were not detectable at low doses.

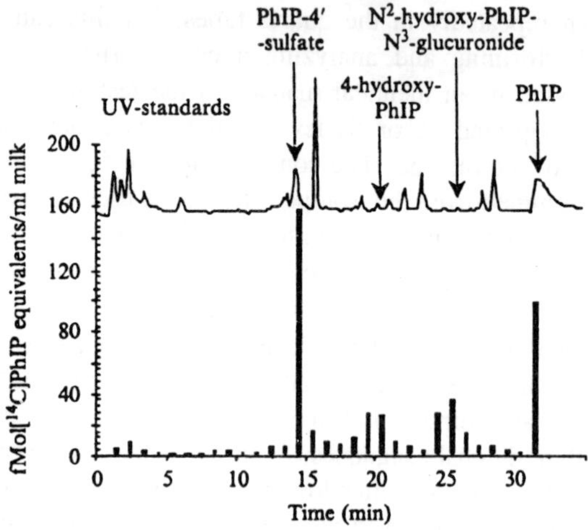


*Figure 8.14 : High-performance liquid chromatography-accelerator mass spectrometry (HPLC-AMS) isotope chromatogram of $^{14}$C-PhIP metabolites detected in breast milk from a female rat administered PhIP at 1000 ng/kg. The top chromatogram shows ultraviolet detection of metabolite standards.*

As stated previously, other separation techniques can be used in conjunction with AMS. Binding of $^{14}$C-labeled molecules to proteins has been quantified and specific protein targets identified, by combining polyacrylamide gel separations (one dimensional or two dimensional) with AMS detection. Typically, gel bands are excised and after drying under vacuum can be directly analyzed, because the polyacrylamide, which is produced from petroleum-based chemicals, contains sufficient carbon (51%) to act as a carrier. The total amount of carrier carbon can be calculated if the volume and composition of the excised gel band is known.

### AMS Detection of Chemically Modified DNA and Protein

AMS has demonstrated great value in determining whether chemicals bind to DNA or protein, forming covalent adducts. This is important because DNA adduct formation is considered an early initiating event in chemical-induced cancers. The presence of protein adducts, which may be formed at higher levels than DNA adducts, indicates a chemical is either itself reactive or can be converted to a reactive metabolite capable of binding to cellular macromolecules. Protein adducts can serve as surrogates for DNA adducts, giving a measure of an individual's level of exposure to the bioactive dose of a compound. Techniques used for detecting adducts must be very sensitive because adducts are usually formed at extremely low levels, particularly in humans exposed to environmental or therapeutic doses of carcinogens or drugs. AMS has been used to detect DNA adducts at levels of 1–10 adducts/$10^{12}$ nucleotides, which is less than 1 modification/cell, following acute and chronic exposure to $^{14}C$-labeled carcinogens. This is at least 100-fold lower than the next most sensitive detection method, the $^{32}P$-postlabeling assay.

One of the first biomedical applications of AMS was to determine whether the compound MeIQx, which is formed in cooked meat, binds to DNA in rodent tissues at low doses and whether the relationship between adduct formation and dose is linear. Since then, this technique has been used to answer similar questions for a number of chemical carcinogens in humans and animal models, including benzene, trichloroethylene, and the breast cancer drugs tamoxifen and toremifene.

In this type of study, DNA and protein are isolated from tissues using standard protocols. For measurement of DNA adduct levels, tissues are lysed and digested using the enzymes proteinase K, RNase T1, and RNase A. The lysate is loaded on to a Qiagen anion exchange column, and the protein and other material, including non-covalently bound labeled compounds and metabolites, are washed from the column, while the DNA is retained. The DNA is then eluted, precipitated, and washed with 70% ethanol before redissolving in a buffer appropriate for AMS analysis. It is important to demonstrate that all non-covalently bound compounds are removed by the isolation and purification procedure employed. For example, DNA could be enzymatically digested to nucleosides and separated by HPLC, ideally using a system that resolves adducted nucleosides from unmodified nucleosides and the free parent compound and metabolites. This method should also remove other potential $^{14}C$-labeled contaminants that may be present, such as residual adducted protein or peptides. AMS analysis of the collected fractions will indicate what percentage of the total $^{14}C$ or $^{3}H$ signal can be attributed to covalently bound adducts. Because this will be compound specific, performing this type of experiment will identify situations in which it may be necessary to incorporate additional DNA purification steps.

To isolate protein (and other acid precipitable macromolecules), homogenized tissues are lysed overnight and then centrifuged. Perchloric acid (70%) is added to the supernatant precipitating the protein. After centrifuging, the resulting protein pellet is washed with 5% perchloric acid, followed by several organic solvents (a 50% methanol solution and a 1:1 mixture of ether:ethanol) to extract any residual non-covalently bound $^{14}C$-labeled compounds. The protein is then redissolved in 0.1 M of potassium hydroxide and aliquots of this solution are submitted for AMS analysis.

AMS can also be used to determine the ability of chemicals to bind to specific proteins, as was demonstrated by Dingley et al. in a study that reported PhIP binds to the blood proteins, hemoglobin, and albumin and to white blood cell DNA in humans administered a dietary relevant dose of [$^{14}C$]-PhIP. Adduct levels were monitored in five subjects over a 24 h period by taking blood samples (30 ml) at various times and separating into plasma, red blood cell, and buffy coat (containing white blood cells and platelets) fractions. Albumin and hemoglobin were isolated using standard protocols, with the addition of an initial dialysis step, and analyzed by AMS. It is crucial when extracting albumin or hemoglobin that each sample of plasma or red blood cell lysate is first dialyzed extensively, in an individual beaker, for 48 h to remove non covalently bound $^{14}C$ labeled compounds and metabolites. This investigation demonstrated the value of AMS in providing information on adduct formation and kinetics in humans, which may be important in assessing an individual's susceptibility to specific carcinogens.

Protein samples, which are usually obtained in higher quantities than DNA, can normally be measured without the addition of carrier. Approximately 3 mg of protein will yield the required 1 mg of carbon, as will 3 mg of DNA when this amount is available, because both are composed of about 30% carbon, although this is dependent on tissue type and species. Smaller quantities can be analyzed (as little as 1 mg of DNA or protein), but there is greater potential for contamination and the sample may not produce sufficient graphite. In situations when the amount of tissue, and therefore, DNA is limited, as is often the case with human studies, the DNA is submitted with carrier. The amount of DNA analyzed in this way can be as little as 1 mg, although this will be influenced by the expected level

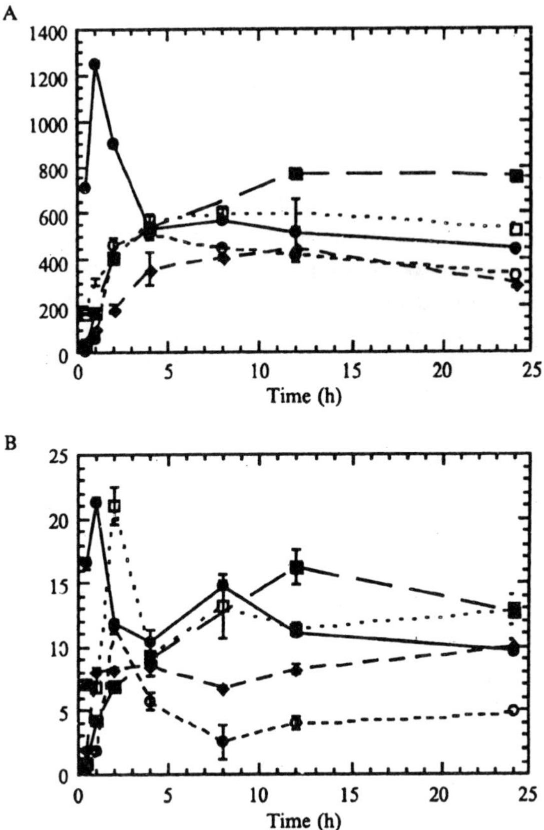


*Figure 8.15 : Binding of PhIP to albumin (A) and hemoglobin (B) isolated from five human subjects administered a single dose of 14C PhIP.*

of adduction, the compound specific activity, and availability. Material analyzed neat does not need to be accurately quantified, because the AMS measurement is a ratio, but for carrier supplemented samples, the precise amount of DNA or protein must be determined. Consequently, when analyzing carrier added samples, the precision and accuracy of the AMS data can be limited by the precision and accuracy of the methods used to quantify the mass of material. Alpha particle energy loss measurement was developed as an accurate way to nondestructively quantify the mass of microgram quantities of macromolecules with high precision before analysis by AMS.

## THE FUTURE

AMS methodology has advanced substantially over the last 10

years, resulting in its now routine use in the measurement of $^{14}C$ and to a lesser extent $^{3}H$ in biological samples. The technique offers unique capabilities in studies where low amounts of radioisotope labeled compounds must be quantified with high precision. These qualities have led to the increasing use of AMS in areas of nutrition, biomarker detection, human metabolism studies, and drug development. Methods for the analysis of additional isotopes are being developed.

Over the coming years, we anticipate AMS applications will expand and become more diverse as smaller cheaper instruments designed specifically for biologists are constructed and the technology becomes more accessible. Furthermore, automation of the sample preparation process will increase sample throughput and could allow for analysis of smaller samples, all of which will help expand AMS into all areas of biological and environmental research.

# Index

## A

## B

## C

## D

## E

## F

## G



## H



## I

## J



## K

## N



## O

## S

## T

## U



## V



## W



## Z

U